UNDERSTANDING PRINCIPIA AND TRACTATUS

Russell and Wittgenstein Revisited

A. P. Rao

Understanding Principia and Tractatus

Russell and Wittgenstein Revisited

A. P. Rao

International Scholars Publications
San Francisco - London - Bethesda
1998

Library of Congress Cataloging-in-Publication Data

Pampapathy Rao, A., 1936-
Understanding Principia and Tractatus : Russell and Wittgenstein revisited / A.P. Rao
p. cm.
Includes bibliographical references and index.
ISBN 1-57309- 099-9 (hc : alk. paper). -- ISBN 1-57309-098-0 (pb : alk. paper)
1. Wittgenstein, Ludwig, 1889-1951. Tractatus logico -philosophicus. 2. Russell, Bertrand, 1872-1970. Principles of mathematics. 3. Logic, Symbolic and mathematical. 4. Language and languages--Philosophy. 5. Mathematics--Philosophy. 6. Description (Philosophy) 7. Identity. I. Title.
B3376.W563T7355 1997
192--dc21 97-7115
CIP

Editorial Inquiries:
International Scholars Publications
7831 Woodmont Avenue, #345
Bethesda, MD 20814

To order: (800) 55-PUBLISH

TO MY WIFE

WHO KEPT STIRRING AN EMPTY POT

WHILE I WAS SPILLING INK

Table of Contents

Preface

Every eagle has its own rock from which he sets off for day long flights and on which he perches after the Sun sets. Similarly a philosopher keeps periodically revisiting the thinkers that influenced him during his initial training. Towards the evening, before settling, the eagle takes a close look at his rock; and towards the close of his career the philosopher does the same. He takes a second look at the philosophers who in one way or the other had been points of reference to him. In my self-training in philosophy I paid greater attention to Russell of the Principia and Wittgenstein of the Tractatus; and ever since, they remained for me paradigm philosophers. Fortuitously, as my subsequent life came to unfold, my relationship with professional philosophy and professional philosophers too turned out to be analogous to theirs. Now towards the evening of my professional life I started looking anew at what interested me most in their philosophical thought. The outcome is these two exercises in understanding their basic contributions to philosophy. I tried here to show how two great minds are wrestling with problems which could not have been posed before they solved them, and how this is indicative of their genius. However, it should not be thought that the fragments of their respective contributions to philosophy which I have considered here exhaust my interest in them; obviously, these are fragments also of what still stands in their contributions untouched by time, and thus retain for them prominent place among contemporary thinkers. But these fragments are so fundamental to their respective philosophies that they would not be worth noticing but for these. In Part One, I pay exclusive attention to 14.01 of the Principia; and the entire Part

Two is devoted to understand 3.203 of the Tractatus. Thus what follows is an unpacking of a definition of Russell, and a remark of Wittgenstein. Nevertheless, there is nothing recondite or pedantic about it. Philosophy - even "pure philosophy" in which hard core logical and metaphysical issues are handled in great detail - need not be dull, boring and unreadable.

The object of concern in Part One is 14.01 of the Principia, namely

$$\{(\mathrm{I}x)\phi x\}\psi(\mathrm{I}x)\phi x =_{\mathrm{df}} (\mathrm{E}y)\{\psi y \,.\, (x)(\phi x \equiv x = y)\}$$

as it can be formulated in a familiar notation. If in this part I confined my discussion to Russell's theory of definite descriptions it is because that though a century is about to pass since it was presented, and though a whole library has been published on it, we still find Russell's problem only in its distorted versions. Far too many problems which others, obviously with his theory in the background, came to be aware of have been imputed to him. His theory has come to be judged and found wanting on the count that it does not incorporate rubrics for solving those problems. In the first chapter of the first part, I paid some attention to the most important contributions to the secondary literature on Russell's theory to impress upon the reader why he should try to understand Russell's theory bracketing that literature. Nevertheless, I took note of the contributions of Strawson, van Fraassen, and Donnellan but only to show that their puzzles, and what they consider to be Russell's problem, are not of Russell's concern. Moving towards isolating Russell's problem I considered several initially unconnected puzzlements of Russell concerning the logical behaviour of the definite article. Setting apart its logical role from its grammatical role in the languages in which it figures, I argued that the problem which sparked off Russell's decade long enquiry arises not only with respect to the languages in which it figures - and thus is not idiosyncratic to English - but in any language in so far it is a vehicle of representation. I considered at length the way in which the definite article functions in the languages in which it figures - specially, 'the' in English - to ascertain whether it behaves as a quantifier as a function, and argued that it behaves like an operator saturating the predicate expression to which it is prefixed, and showed how noticing this constitutes Russell's basic insight on which is based his theory of descriptions. Further, I argued that given this it should be possible to capture the behaviour of the definite article within the framework of quantification theory as developed in the Principia.

In the second chapter I sketched the problems which mathematicians from Weirstrass onwards - specially, Dedekind, Frege, and Peano - faced while attempting a definition of the concept of an integer, and how Russell's study of their work on foundational problems sparked off some fresh thoughts which finally culminated in the Principia. (This sketch however is not historical in orientation; it is only conceptual, and is focused on only one issue, namely how these foundational studies influenced the development of Russell's theory of definite descriptions.) I tried to show how responding to the work of these mathematicians Russell came to take a stand on several issues, for example, whether mathematical object theory is to be preferred, or mathematical property theory is to be accepted, whether proper names and definite descriptions are of the same semantic category or not, and whether mathematical and philosophical definitions are of the same nature. I traced how, while thinking about these issues, Russell came to note the essential nature of a mathematical definition. Paying attention to his thinking prior to the Principles, I showed how Russell came to characterize a mathematical definition in his Principles as follows:

> " ... given any set of (logico-mathematical) notions, a (logico-mathematical) term is definable by means of these notions, when and only when, it is the only term having to certain of these notions a certain relation which itself is one of the said notions."

I explained how Russell exploited this insight into the nature of a mathematical definition to handle two sets of issues which confronted him in two disparate areas of his interest. First, he tried to bury Meinongianism - which is all pervasive in the Principles - by defining 'the' in terms of 'all', 'some', 'if-then', etc., and secondly, he saw a way out of the limitations of Peano's definition of the number concept (which I spelled out in some detail) to be able to carry out his logicist program. I tried to show how in these attempts Russell faced new puzzlements, specially about identity, for example whether it was a relation or a function, and if it is a relation, whether the relata are terms or objects. Thus, to be able to logico-mathematically define descriptions, I argued, he had to develop his logic of identity. This became necessary as firstly Peano's tool box did not include a logic of relations, and secondly definite descriptions, as Russell noticed, involved identity. But Russell's definition of identity required the axiom of reducibility; thus he came to think of the Principia order, namely axiom of reducibility, theory of identity, and contextual definition of definite descriptions. I plotted this order without allowing history, text, and technicalities dominate the argument and exposition, and highlighting conceptual links between them and the philosophical problems those links involve.

In the third chapter, in addition to showing how Russell connected all his earlier insights to arrive at the theory if descriptions, and attending to a clarification of the entire conceptual structure which Russell forged to state his theory, I stated his theory as he presented it in the Principia, this is to say that in its full-blown version with all the prerequisites like the axiom of reducibility and identity. In this respect I departed from the accepted tradition of paying exclusive attention to "On Denoting". I showed that the traditional statements of Russell's theory result in serious misunderstandings, a case in point being Strawson on Russell. I departed from the accepted tradition and followed Russell on other counts too, for instance in refusing to take 14.01 of the

Principia as a philosophical definition, specially of the Moorean kind. Among the clarifications I offered here there is an extensive discussion of the notion of presupposition. It was pointed out, not too infrequently - specially by Strawson and van Fraassen - that Russell lacked this notion. I showed, with citations from the Principia, that this is a clear misrepresentation of Russell, and that he in fact considered it at length and tried to highlight the presuppositions involved in the use of definite descriptions. I clarified the sense in which definite descriptions are said to be incomplete symbols, and explained their role in representation., Another important issue to which I paid sufficient attention is the relationship between the definiendum and the definiens in Russell's contextual definition. Having cleared several confusions concerning this relationship, which arose due to taking 14.01 of the Principia - contra Russell's intention - as a philosophical definition, I pointed out that contextual definitions should not be seen in terms of the received conceptual pair "definiendum - definiens," but in terms of the conceptual triad "definitum - definiendum - definiens." Further, as Russell's definition requires an understanding of the notions of existence and individuation - apart from that of identity, obviously - I explicated these, and explained how Russell interlinks them. It could be that these and some other clarifications that I tried to provide in this chapter needed to be attended to even at this stage as till now nobody worked out a full-length coverage on Russell's theory of descriptions focusing attention on 14.01 of the Principia and its role in the kind of the system projected in it.

With this in the background, I moved to the fourth chapter to consider the criticism of, and alternatives to Russell's theory suggested by David Kaplan, Dana Scott, and Abraham Robinson. These alternatives to his theory are claimed to be programs within Russell's own framework; they are supposed to be in the spirit of his theory. I showed that they are not what they are claimed or supposed to be. My evaluation of these alternatives is based on two principles of philosophical criticism. The first of these is that one should not attack the opponent except at his strongest point; this I call the principle of fairness. The second is that one should not shift the problematic - this is the principle of pertinence. I showed that the alternatives suggested or sketched by Russell's critics fare badly on these counts. As my defense of Russell against these critics required it, I offered here explanations concerning several important strategies and concepts, for example elimination of descriptive phrases using induction, semantical and syntactical features of sentences containing multiple descriptive phrases, and the notions of logical simplicity, and termial and sentential substitution. I showed how anomalies crop in each of these alternative systems, and argued out why none of them can be preferred over Russell's theory.

In the fifth chapter I considered the alternatives worked out by Karel Lambert and Jaakko Hintikka, and argued that these are not serious contenders on the count of the criteria of criticism spelled out and defended earlier. I argued that in rejecting one half of Russell's definition, they shift not only the problematic but also the conceptual basis of Russell's theory itself. I showed how their systems are founded on an altogether different relationship between existence, identity and individuation. In fact, it turned out that they have a concept of logic which is totally different from the one to which Russell subscribed. Thus it emerged that there is no plausible Russellian alternative to Russell's theory, and that non-Russellian alternatives are as plausible or implausible as Russell's theory. The upshot of this effort is that Russell's theory can cope with the problems it was intended to handle, and thus very much a live contender rather than a historical curiosity.

Finally, in the Appendix, I set aright some serious misconceptions about Russell's concept of a mathematical definition which fueled equally serious misunderstandings about the contextual definition of a definite description.

Part Two is devoted to understand 3.203 of the Tractatus, thus the pithy remark:

> "Der Name bedeutet den gegenstand. Der Gegenstand ist seine Bedeutung. ("A" ist dasselbe zeichen wie "A".)"

What Wittgenstein is saying here, though not why he is saying it, should have been clear to everyone but for the grooves of understanding which the British interpreters and translators of the Tractatus got into; after all, in philosophy a translation is an interpretation. In any case, as I am not as blessed as they are, neither Wittgenstein himself, nor angels or tutors on his behalf, whispered to me what he really meant, I take it that in 3.203 Wittgenstein is remarking that

> A name refers to an object. The object is its referent. ('A' is the same sign as 'A'.)

and then try to understand why this happens to be the case, or how this obtains precisely in this way. The whole of Part Two is geared towards this end; and from this exercise the close relationship between the two parts of this volume comes to the fore. Obviously, the point of reference for Russell as well as for Wittgenstein is the same, namely the standard quantification theory presented in the Principia. Both of them accept the necessity and the sufficiency of this theory for the purpose of representing the world in language. If they differ, they differ on issues pertaining to the limits and the presuppositions and the tacit assumptions of this theory, specially whether the notion of identity is internal or extrinsic to it. Russell thought that it was extrinsic, so tried to extend it to envelop identity too; and Wittgenstein thought that it was intrinsic to it, and thus tried to reinterpret its representational mechanism. But this small issue is so fundamental that any stand on it has its own far-reaching consequences. In the concept of identity, fortunately or unfortunately, philosophers have to them something a stand on which determines what one has to hold on everything else, and also constrains whatever one has to maintain on anything else. Here is a message to students of philosophy, namely: take care of identity, it will take care of your logic, your metaphysique, in short, yourself! Thus it is not surprising to find that Russell and Wittgenstein with their respective distinctly different conceptions of identity were forced to project alternative visions of the quantificational mechanism, and yet articulate their respective visions using the same conceptual kit - and articulate them, as if to further complicate the matter for us, mutually interacting.

Though the Tractatus begins with what apparently sounds like a creation myth, its author was not thinking de novo; he was spinning his vision of language and of representing the world in it in the context of, and as an alternative to, the views of Frege and Russell. As 3.203 is not a stray remark but an integral part of a theory of representation, in the first chapter, I stated succinctly and without resorting to punditry, the Fregean and Russellian positions on the issues involved in representation, and spelled out in brief Wittgenstein's alternative clarifying the entire conceptual apparatus he erected to handle the problems that needed to be solved in order to be able to present a plausible theory of representation. I attempted to do this ignoring the secondary literature which is as vast as it is misleading. But this exercise in gaining intelligibility was restricted in this chapter to the principal part of 3.203; here I did not pay any attention to the parenthetical part of it. The results arrived at in this chapter naturally lead to some puzzles. What, after all, does the activity of coming to know that names refer to objects and why this should be so amount to! What is it that one is trying to do when he says that objects are referents of names?

In the second chapter, I attempted to characterize the kind of activity with which I was concerned in the first chapter; and for this purpose I smuggle into my discussion Wittgenstein's notion of natural history. In order to explicate this notion I decoded paragraph 415 of the Investigations, or - as it happens to be the same as - paragraph 141 of Remarks on the Foundations of Mathematics. Having clarified what Wittgenstein means by natural history and having shown how he uses it to handle a host of important issues in philosophy,

I tried to show that 3.203 is geared towards providing a natural history of naming. I do not claim that Wittgenstein during his second innings realized that this is what he did in the Tractatus, or that he intended 3.203 in this way. There is no textual evidence to that effect, except the prefatory remark in the Investigations that it should be read juxtaposing it with the Tractatus. But when an explication of the concept of natural history from the Investigations is juxtaposed with an explication of a concept of a name in the Tractatus, it turns out that they are related in the way that I tried to see them to be related.

But are names indispensable for representing the world? Can a language succeed in representing the world without containing names in its vocabulary? We may recollect that Russell too was puzzled as to whether definite descriptions, and even proper names, are necessary for this purpose. To be able to ascertain Wittgenstein's position on this issue, it became necessary to understand his remark in the Tractatus that the world can be captured in terms of generalized sentences. So I devoted the third chapter to paraphrase in great detail 5.526 of the Tractatus, and then sketched the way in which Wittgenstein fixes the role of names in representation. Further, this brought about some intelligibility concerning sentences of the form Fa by showing that they mean the same as sentences of the form (Ex)(Fa . x = a). This means that identity is involved not only in the use of definite descriptions as Russell noted, but even in the use of proper names as Wittgenstein noticed.

In the fourth chapter, which is much larger than the first three, I returned to the parenthetical part of 3.203 and tried to bring out its import by relating it to 5.533 and remarks related to it. This resulted in an elaborate, if not exhaustive, discussion on Wittgenstein's views on identity, specially his criticism of Russell's theory of identity in the Principia, and his response to Ramsey's suggestions towards a revision of the Principia to redeem it from the theory of identity which is based on the axiom of reducibility which in no sense is logical in its nature. Having shown how Wittgenstein thought that as quantification theory generates identity from within, provided its syntax is taken care off, there will be no need to extend quantification theory, à la Russell, to envelop identity. I argued that in the parenthetical part of 3.203 Wittgenstein is encoding the principle of constancy of reference, and that this principle is in effect equivalent to the law of reflexivity of identity.

I intended this part as a clarification of Wittgenstein's thoughts; it turned out to be an elaboration and embellishment of his themes. Further, these elaborations - which are not to be treated as hermeneutical - are elaborations of Wittgenstein's views on some central logico-philosophical issues in the light of his apparently strange epigrams and pithy remarks not only in the Investigations, but also in Note Books and even Culture and Value. They show that they all - his professional philosophy and his proverbial pronouncements - together project one great architectonic which he worked out in parts for the like-minded "scattered across the globe."

A.P.R

Acknowledgment

I am deeply beholden to Dr. Robert West, and Dr. R.S. Bhatnagar. But for the Job-like patience of Dr. West towards my multiple failings, and the ungrudging help of Dr. Bhatnagar in several matters, the following pages would have been blank; and they would have irritated the reader but for Ms. Anne-Elisabeth Mercier who prepared the final script.

A.P. Rao

PART ONE:

On 14.01 of the Principia

<u>Chapter One : The Problem After One Hundred Years</u>

A quick glance at the voluminous literature on phrases with the definite article as their prenex - for example, 'the author of Waverley', 'the horse caught by Bellerophon', and 'the substance of the world' - which appeared since the turn of this century reveals how everyone who contributed to this still growing corpus was aware of the anomalies in and the inadequacies of all the theories and explanations offered by everyone else and was eager to avoid them while being indifferent to the consequences and presuppositions of his own theory or conjuncture. The disagreements between these contributors in general, and the disagreements between the trendsetters among them in particular, seem to be as much about what constitutes a solution to the problem as they were about what indeed the problem was. It would be far from truth to say that all of them were after a solution to the same problem. Scanning this literature one can come across instances of some of these contributors brushing aside as marginal to the issue what other have identified as central to it. In some cases the problem, or a bunch of problems, seems to have been overdetermined to the point of its growing a resistance to any solution. And occasionally margins of precision seem to have been unwarrantedly widened to allow soft solutions gain a semblance of success, or to make way for high-energy techniques coercing the uninitiated to believe that the problem has been finally solved. It is also the case that not too infrequently, searching for the historical hair-roots of some hard perennials, the whole garden was dug out unsettling even the well-established blooming bushes. Worse is that a large number of these contributions appear to be the results of attempts not to handle the encountered issue, but to contrive a solution subsuming the issue as well as the solution under a belief or a design which is extraneous to the area under investigation. Perhaps with the exception of modal logic, no other area of logic in which philosophers take some interest shows how "theologicians" - this is to say that logicians with theological interests - have highjacked logico-mathematical investigations to save or serve God than the enquiry into the nature of descriptive phrases. Thus it is not surprising if in such a situation a large number of enquiries concerning descriptions turned out to be attempts to save examples[1] which happen to be supreme fictions like 'the present King of France', 'virgin John's children', 'Pegasus', and 'Santa Claus'.

Further, enquiries into so cluttered an area is likely to perpetuate the very same vices or worse mistakes unless far too many arrogant assumptions resulting in confident conclusions are sacrificed in favour of asking ignorant questions requiring simple but sure answers. To put it differently, at this stage of the enquiry concerning definite descriptions it may turn out to be fruitful to follow Mach's methodological precept and bracket much of what has been done and is claimed to have been achieved till now, and start isolating the problem in its simplicity and clarity. This does not however mean that we should ignore all the extant work and begin de novo; that surely will be another arrogant act leading us to a repetition of the earlier mistakes. Conscious originality is the breeding ground for secondhand errors. But it seems to be reasonable to ignore an inter-connected bunch of writings on the issue if it happens to be based on an assumption, or if it incorporates a strategy to handle the problem, which has been shown to yield results that fail to meet the demands of logic and legitimate expectations about logic, or else is incompatible with well-established results in logic. I trust that this is not demanding much unless of course there are compelling independent reasons for abandoning the received conception of logic itself or else to replace the received body of logic by a new system with which the consequences of the assumption or the strategy in question are compatible. On the other hand to demand a revision in the received concept of logic so that those consequences may be accommodated and legitimized is to sacrifice everything to save an example, which if not worse is tantamount to committing a fallacy of enquiry, at least. In fact most of the published work in this area creates the impression that it was carried out almost to vindicate the Wittgensteinian remark that on an unbalanced diet of examples our thinking develops deficiency diseases. Yet, giving an example to illustrate the point at hand may be in order; and the example I offer, with all the risks involved in it, is a double ended one. It is intended not only to illustrate the point but also at the same time to indicate the requirements to follow and evaluate the present enquiry, this is to say that how this enquiry assumes on the part of the reader some familiarity with the literature to which I have been referring, and his or her willingness to have a second look at it with an open mind.

It is a commonplace to all contemporary philosophers concerned with definite descriptions that Strawson (1) found fault with the way in which Russell[2] in his (2) analyzed them and that with that the enquiry initiated by him took a different and perhaps interesting turn. It is well recognized that there are several enticing and much debatable issues involved in the twist given to the investigation by Strawson. Some of these are whether Russell and Strawson both are talking about the same thing, whether both of them are equally concerned about the distinctions which the latter draws between sentences, their respective uses and utterances - specially not in what either of them say about these distinctions but in what they do with these - and more importantly whether both of them are raising issues at the same level, be it syntactical, or semantic, or else of pragmatics. It is also doubtful whether both of them are subscribing to precisely the same relationship between terms and their presumed referents on the one hand, and the same conceptual relationship between terms, predication, and truth-value on the other. But to initiate the projected enquiry I shall ignore these and the related issues, and be content with noting that Strawson was dissatisfied with Russell's analysis of definite descriptions on the ground that it does not pay any attention to the notion of presupposition. The differences between the respective positions of Russell and Strawson developed into a debate between two voluntary armies - at least, there are no mercenaries on the side of Russell - each carrying the flag of one of them, until it naturally became dull and van Fraassen (1 and 2) with his attempt to formally capture the notion of presupposition deploying the strategy of supervaluations based on the notion of truth-value gaps opened a new front. van Fraassen's intelligent exploitation of the notion of presupposition - whether it really was due to Frege, is an issue that needs to be settled by those who are better acquainted with his work - was believed to have provided the rubrics required to link problems concerning definite descriptions and problems that arise in connection with other kind of terms and thus relate two enquiries which hitherto were carried out in isolation[3]. Further this coupling of two sets of problems - variantly put, this unification of the theory of descriptions and free logic - to arrive at a unified system of singular terms, be they definite descriptions or proper names, is claimed to have been achieved without sacrificing any of the received laws of logic, thus within the confines of the received conception of logic. But unfortunately this program ended with some unpleasant results; in addition, it took recourse to rejecting the principle of bivalence and resorting to a cavalier treatment of sentences containing identity and singular terms. Even if these are treated as innocuous, and even if one does not harp too much on the view that logic is a set of truths to which van Fraassen's program is committed, it would be difficult for anyone to ignore the results that Woodruff (1) established. He proved that

(i) the set of supervaluationally valid arguments is not recursively enumerable[4], and

(ii) supervaluational semantics is not compact, and thus the strong completeness theorem will not be available for it.

There are no compelling reasons for going along with van Fraassen and hold that, in view of these proof-theoretic and model-theoretic results, we need to revise our received conception of logic which with its standard interpretation has the properties of recursive enumerability and compactness of the sets in question. In so far as the received conception is concerned, the two notions, namely that logic is set of valid formulas and that it is a set of valid argument-patterns coincide. Postponing details to Chapter Five we may note that van Fraassen's conception of logic is at variance with the received conception - despite his claims, that is - in another and an equally important sense. The received conception of a truth-functional connective breaks down under supervaluational approach. For instance, the wedge which is a truth-functional connective under classical valuations ceases to be so under supervaluations, for though 'Pegasus = Pegasus' is undecided, '(Pegasus = Pegasus) ∨ (~ Pegasus = Pegasus)' will come to be assigned, under supervaluations, the truth value Truth. Thus the wedge as in the context of classical valuations cannot be said to be the same as it occurs in the context of supervaluations.

The central thrust of Strawson's argument however is not that the received conception of logic as it figures in the Principia stands in need of a revision or replacement. His point on the other hand is much more cynical and despairing depending upon the way one looks at it. It is that inference patterns in ordinary language are inexact and that the received language "has no exact logic."[5] As such either replacing the received logic, or revising the earlier rejected Aristotelian or Port-Royal logic is of no consequence. Consistency demands Strawson to dismiss van Fraassen's attempt to formalize his notion of presupposition precisely for the same reason; if he did not do that it could be due to the fact that the consistency which requires it is exact consistency which, according to him, is alien to the received language[6]. In any case van Fraassen has no axe to grind as far as the set of logical truths accepted in the received system of logic is concerned. The source of irritation for him is the semantics provided for that set; to be precise, his gripe is at the style of the received semantics not being ontologically neutral. But on that count his reinterpretation does not fare any better. To withhold truth-value assignment to a sentence whose subject term is non-denoting may be, as van Fraassen believes it to be, what it is to be ontologically neutral; but it is also being cavalier in and to ontology. More importantly, this cavalier treatment of ontological issues is made possible by some firm epistemological prejudices. Thus the choice is between the ontological Scylla and the epistemological Charbydis. Van Fraassen's rejection of the principle of bivalence, and his acceptance of truth-value gaps show his nominalistic bias - anti-realistic prejudice, at least. Epistemological realism[7], as Dummett has argued at several places all these years, amounts to that each member of the set of sentences is either true or false[8]. This should indicate how many of our logical differences are rooted in logic, and how many of them are located in metaphysics. That apart, is it not rational to require a formulation of logic to preserve the two properties, namely recursive enumerability of truths, and strong completeness? Should we not then ignore the mountain of writings based on supervaluations and truth-value gaps, and rescue the problem from being buried under it! It seems that it is time to realize that van Fraassen has misdirected the program to bylanes; and to drive it to the main path again it first needs to be recognized that either he misrepresented the Strawsonian notion of presupposition or else his formal characterization of the notion is not all that is given to it. Surely, Strawson would be the last philosopher to savour all that flows from Fraassen-gaps[9]. Thus it seems to be fair to bracket van Fraassen's intervention in the debate and return to the problem - whatever it might be - where Strawson left it. There is nothing wrong in doing so as the industry of philosophers need not necessarily produce progressive illumination. But before I do that, I may draw the attention of the reader to at least one more product of that industry to show how I intend to keep aside another set of contributions which seem to have led us astray ad distanced us from the problem.

It is a tacit assumption of philosophical polemics, thus a large part of post-war professional philosophy, that any two mutually opposing theories are rooted in the same set of errors. So the third invariably starts lumping them together. But the very assumption preempts philosophical progress from being epistemically incremental. Donnellan's much cited paper is a case in point. Donnellan (1) starts bracketing Strawson (1) and Russell (2) on the count that both of them are reference-theories, and neither of them takes into account the attributive role of definite descriptions. He has nothing to add to what he understands Russell to be saying on definite descriptions in so far as they are used referentially; and what he understands Russell to be saying is only a part of what Russell did say, but we shall return to this a little later. If Donnellan were to hold that the issue handled by Russell and Strawson is a pseudo-issue, and that the real problem is the one which he is posing, then, obviously, his explanations can be tested against the goals he has on his mind; and if he were to claim that Russell and Strawson raised only a partial problem and that he is attempting to pose it in its entirety, then we can ask about the relevance of his explanation in providing solutions to the difficulties Russell faced[10]. But neither Donnellan himself nor his followers took his contribution in either of these ways. Their writings create the impression that they believe that Russell and Strawson attacked the problem with an inadequate kit of tools and that Donnellan produced a better tool-kit. This is to say that they believe that they were not shifting the problem but were only attempting at a new solution to the old problem equipped with technically better tools. Then the appropriateness of his solution to his tools need to be assessed in relation to the conceptual context in which the problem has been posed by Russell. Obviously the background concepts and the perceived or the imposed relationships between them form the coordinates which individuate the problem. Thus if we can show that the conceptual battery of Donnellan is either incompatible with, or is a misfit in, the conceptual background of Russell's problem, we will have shown that Donnellan is attempting to solve his own problem to his satisfaction and is not attacking Russell's problem from a new angle. Deferring details about the conceptual context of Russell's puzzlement concerning definite descriptions to a later stage in our enquiry, at this stage we may consider Donnellan's discoveries taking into account only some salient features of that context, in fact those features which are unlikely to be sources of a disagreement among Russell scholars and hence do not stand in need of immediate explanations and clarifications.

It is common knowledge that departing from the Fregean practice of keeping proper names and definite descriptions in the same semantic basket - Frege lumped them together under the category of *nomen* - Russell held that there is a categorical distinction between them. Of course in a canonical language[11] like that of the Principia[12] both of them can be defined away (?) in terms of the quantifiers variables and identity; but in a language which has proper names among its primitive terms - for instance, ordinary language, as it is usually believed - there will be an irreducible syntactical difference between proper names and definite descriptions, and this will be the case even if both of them are made to behave alike semantically. But what obtains with Donnellan's conceptual pair - namely, attributive and referential uses of definite descriptions - is that it vitiates the distinction between proper names and definite descriptions; it accomplishes this at least in so far as these constitute the subject matter of an enquiry. After all, we do use proper names and definite descriptions referentially as well as attributively, and use them in both these ways not only figuratively but literally as well. We spin any number of indicative sentences in which proper names are used attributively; for example, asked why I keep working even on public holidays I might say: "I am no Gauss, so I need to work like a mule." Note the way in which 'Gauss' figures in 'I am no Gauss'. If one is not happy with this example because an indexical expression figures in it, he may consider the following sentence: 'Anyone who is not a Gauss will have to work like a mule'. The quantifier and the indefinite article occurring in this sentence should be sufficient to convince any proficient user of English that it is an indicative sentence in which a proper name figures attributively, and that this use of the proper name here is not enveloped by the figure of speech that figures in the latter part of the sentence[13]. Thus, in so far as the possible ways in which proper names can be used by a proficient user of a language is concerned, they cannot be set apart from definite descriptions. This means that Donnellan's invention decomposes without residue a crucial distinction on which Russell's theory of descriptions is based. As it cannot be imagined even that Donnellan is unaware of this, it may have to be concluded that he is concerned with his own and not with Russell's puzzles. Worse is that Donnellan, not unknowingly, is putting the cart before the horse. Whether a specific use of a sentence is an instance of its normal use or else is a case of its misuse can be decided upon only when semantic matters concerning it are settled antecedently and independent of its use by anyone. If not, we will be back at the egocentric predicament - which Russell sarcastically or in a mood of exasperation dismissed in (1), and - whose essence may be captured by the much mentioned Lewis Caroll joke: the issue is not what the words mean, but who the master was. As far as this issue is concerned Strawson and Donnellan are in the same camp; neither of them take into account the knowledge we gained from the science of linguistics. They ignore in particular the fact that syntax, semantics, and pragmatics constitute a nested triad, and that as such settling syntactical issues is necessary for moving to semantics as much as settling syntactical and semantic issues is a prerequisite for being able to tackle issues in pragmatics. I am not suggesting that philosophy is not an autonomous discipline; but I do believe that it is desirable to pursue it in full view of what is happening in other empirical sciences. In any case it is important to note in the present context that a given philosophical problem concerning language it first needs to be placed in one of these subdisciplines of the science of linguistics; its solution requires a knowledge of the level at which it will have to be solved. More often, mistakes are committed in philosophy by misplacing the problem and thereby either overdetermining or underdetermining it either to face despair or to indulge in irrelevancies. Russell (11) made clear what his intentions were when he posed the problem. He was emphatic that he did not intend to do pragmatics, or as he put it, he was not interested in the "psychology" of users. In fact quite early in his long pursuit of the problem he remarked in (8)

> "there is a sense in which we denote, when we point or describe, or employ words as symbols for concepts; this, however, is not the sense that I wish to discuss."[14]

Nor did Russell aim at doing semantics per se; to put it anachronistically, echoing Wittgenstein of the Tractatus who was influenced by his work of the first decade of this century, Russell remarked:

> "In philosophy, it is syntax, even more than vocabulary that needs to be corrected."

In view of this we may safely assume that what Donnellan and his followers have to say may be important in itself, but certainly is not relevant to Russell's issue. Then it is desirable that we should return to the problem where Strawson left it, rather to the point where he joined the search for a solution to it.

But before I return to the Strawsonian point in the enquiry, intending not to give anyone a chance to be appalled by my ignorance of, and not to provide an occasion for being accused of ignoring Kripke's and Kripkean explanations, I may briefly mention my reasons for bypassing them. Tharp (1) has shown that from what Kripke has to say in his classic (1) the following theses can be derived.

(i) Every truth is a priori equivalent to a necessary truth.

(ii) Every truth is necessarily equivalent to an a priori truth.

(iii) Every truth is a priori equivalent to a contingent truth.

It was reported that when Tharp made his results public everyone except Kripke was surprised, and that he was not surprised as he knew all along that his position leads to these consequences[15]. This is perhaps beside the point and hence should not matter much. What should matter is that these Tharp-theorems in metaphysics are highly dubious and have been much contested before as well as after Kripke (1). Any theory or explanation which leads to such consequences should be suspected of serious conceptual confusions and fusions. Thus we may return to Strawson.

Russell's response to Strawson in his (11) no doubt betrays his prejudice against (not ordinary language, but) ordinary language philosophy and his habitual sarcastic summary dismissal of what he dislikes at the moment, rather than show his passion for serious engagement about philosophy of language - he surely had a lot of that passion - or his willingness to enter into a detailed discussion. In any case Russell's reply amounts to saying that Strawson's criticism of his theory of descriptions was tangential. No wonder if since then it has come to be believed that they were addressing themselves to different problems, at least that they were talking on different wavelengths. And - this is rather surprising - that a large body of literature as the one devoted to the issues involved in this reply-rejoinder game nobody seems to have noticed the methodological affinities between Russell and Strawson. More important is that it slipped from the attention of either of them. There are of course personal reasons - reasons of personality, that is - for Russell's not noticing it, for not admitting it even if he noticed it; but this is not the case with Strawson. The responsibility for Strawson's criticism being tangential to Russell's central insights is not with Strawson; Russell has only himself to blame for it because he himself indulged, either inadvertently or in haste, in some tangential talk concerning his fecund discovery with far too many far-reaching logico-philosophical consequences[16]. Though Strawson did not see the affair in this way, it will not be difficult to show that he is not as original as he appears to be taking himself to be, because what he has to say on the issue comes out as a generalization of what Russell did, and say that he did. Though when Wittgenstein, in the early twenties, started articulating the thesis that meaning is use, Russell judged him to have gone senile, he himself subscribed to that thesis two decades earlier. Writing of numbers, Russell then argued at several places that for a proper understanding of the concept of number, the laws of arithmetic alone are insufficient as it requires an understanding of numerical discourse figuring as a part of discourse about other matters, this is to say that it requires an understanding of the application or the use of the laws of arithmetic. Strawson's point - to put it slightly different from the way he formulates it - is that this precisely is the way in which definite descriptions are to be understood. No Herculean effort is needed to show that what Strawson has to say in his (1) is essentially a formulation of what Russell did say in his (2) about the puzzlement of George III, and what he said about what he did towards exiling the then King of France. Isolating what Russell did from what he said that he did, we may note that Russell may have said that he was not exploring the psychology of the users; but then there are innumerable passages in his writings which undermine one's confidence in what he said. I am not suggesting that this is all that is given to Russell's theory of descriptions. There is an important part of his theory and his specification of it which is distinct and is distinguishable from his indifferent talk about it in the spirit of popular science, specially in his formulation and explanation of it in the Principia; and I shall return to it in the next chapter. All that I intend to point out at this stage is that a part of Russell's talk about his theory of descriptions is a perfect fit for the Strawsonian strategy of handling the issue as he understood it. Again I am not suggesting that Strawson was aware of all this in 1950; and I wonder whether he is aware of it even now. My point is neither historical nor biographical. I am only pointing out a conceptual relationship between Russell's talk about his theory and Strawson's criticism of it. Very often the apparently different philosophical vermins are only mutants of common stock, and recognizing this helps us to understand Russell better, specially why he was so touchy in his reply to Strawson. Russell must have realized that Strawson's criticism of his theory revokes the memory of the peripheral arguments which abound in his writings. It was his infirmity that he could not cheerfully bear this feeling; so even in his classic (2) he imputed some such arguments to Meinong and some others to Frege. And he is adept at imputing his earlier erroneous views to others and then refuting them by resorting to attacking the scapegoats[17]. Russell did subsequently realize his mistake and confess in private to Quine. I may as well quote in part his communication to Quine from his (13) Volume 2.

> "In reading you", Russell wrote to Quine, "I was struck by the fact that, in my work, I was always being influenced by extraneous philosophical considerations. Take for example descriptions. I was interested in 'Scott is the author of Waverley' and not only in the descriptive functions of PM[18]. If you look up Meinong's work, you will see the sort of fallacies I wanted to avoid, the same applies to ontological arguments..."

Before decoding this passage to exhibit its intended message, let me first try to keep the whole passage in a perspective. It was written prior to Strawson (1), thus much before Russell (11). Given this it should be easy to pin down the reason for Russell's finding Strawson (1) annoying, if not disturbing. Russell must have felt that Strawson was responding to his "extraneous philosophical considerations," actually considerations which are extraneous even to philosophy, as Russell understood it and as we shall show soon. He must have thought that Strawson confused or conflated those considerations with his distinct contribution concerning "descriptive functions of PM." Russell must have felt that with this confusion Strawson was discrediting him before the philosophical community which by then recognized his theory as a "paradigm of philosophical analysis," to purloin a limpid phrase of Ramsey. Russell is partly justified for, paradoxically, the only work of his which Strawson uses, this is to say that cites and quotes from, is the Principia; two other works, namely Enquiry Into Meaning And Truth, and Human Knowledge Its Scope And Limits are only mentioned and that too in the polemical part of the paper. Russell's (2) surprisingly does not figure at all, though the whole article is under its shadow[19].

To consider the content of the passage quoted above let us for the present ignore the reference to Meinong and the ontological argument, and pay attention to the remaining part of it. What it clearly indicates is that Russell came to realize that (i) there is a philosophical problem, or that there is a bunch of philosophical problems, concerning definite descriptions as they occur in daily discourse, (ii) that there is a logical problem relating to the descriptive functions as they occur in the formal language of the Principia, and (iii) that instead of being monogamously married to the latter he ran after the former too. And as in his life so in his thought the two ladies turned out to be siblings. This is what he intended to communicate in the passage but did not do so explicitly because the correspondent too knows it equally well[20]. That philosophical problems concerning definite descriptions can be reduced to logical problems concerning descriptive functions, or that the former can be solved in terms of the solutions to the latter was integral to Russell's outlook, and was buttressed by his belief that the canonical language developed in the Principia exhaustively exhibits the structure of descriptive language as such. The most important and the lasting contribution of Frege, Russell and Wittgenstein[21] to philosophy is in their recognition that seeing philosophical problems as logical ones is an important step forward, bringing with it a right attitude and a right method to move towards their solution. Thus what Russell was saying amounts to that there is a philosophical problem concerning 'the author of Waverley' and that it can be handled only in terms of the solution to the logical problem concerning say $(\iota x)\phi$ in the language of the Principia, and that problems which cannot be handled in this way are not guenine problems at all. They are pseudo-problems, or else are ones which may have been occasioned by the idiosyncrasies of the language in question; they may have issued out the quirks of the grammar of that language. It is certain that not all problems that arise in relation to the definite article are philosophical; nor is it the case that all of them are rooted in logic. Some of those problems can be located deep in the peculiarities of a given language, and they will have to be handled in terms of the grammar of that language. Consider, for example, a few passages from German philosophical literature; it would be difficult to tell whether one was cracking a joke or was only uttering a truism if he were to say that the more frequent the use of 'der', 'die' and 'das' the more profound the philosophy that was being propounded. The English, on the other hand, do not have much respect either for philosophers or for profundities. So they usually translate German sentences beginning with a definite article by an indefinite one, or even without an article depending upon the proficiency and the aural sensitivity of the translator. In the English philosophical literature, the choice between the definite article and the indefinite one, in so far as only philosophical interests are taken into account, is only occasionally significant. But on these rare occasions the occurrences of the definite articles in German sentences are rendered into English by definite articles only. For example, Hegel's sentences beginning with 'Der Staat' - most of them at least - cannot be rendered into English, without being unfair to him and misrepresenting his intentions, to result in sentences with the prenex "A State'. Had those German sentences been so rendered into English, or had it been possible to render them so into English, Russell in all probability - perhaps coaxed by Mactagaart - would have remained a Hegelian, at least for some more time than has been the case. If he attacked Hegel - and he attacked him bitterly - it was because he was provoked by, as he himself remarked, Hegel's "habit of speaking of 'the State', as if there were only one Global state"[22], which was factually wrong.

This naturally leads one to a point where he feels the necessity to have a handy criterion which can be used in setting apart philosophically significant uses of the definite article from its uses which are not so. Obviously this criterion needs to be one that enables him to set them apart not only in a given language but in any language whatsoever. The import of this statement ought to be evident; still, I would try to draw it out. That a given language does not contain the definite article - or that it does not have articles at all - does not mean that philosophical issues concerning the definite article do not arise in relation to that language; otherwise as an Indian I should not have bothered about theories of descriptions. In fact those problems become more interesting and challenging in relation to such languages because if an occurrence of the definite article is philosophically significant, in virtue of its not being idiosyncratic to the language used in that context, it should be translatable into all other languages including the one in which articles do not figure at all. Philosophical problems have, or should have, this kind of and this much of generality built into them, which is why philosophical problems about language are not problems about a specific language. This means that a given language either has in it the definite article, or else has its surrogates or devices which in effect are equivalent to it such that what is intended to be expressed by a philosophically significant use of the definite article could in fact be expressed in it. The surrogates or the equivalent devices are sufficient to accomplish this task efficiently. After all, all the natural languages have the same expressive power, and this power is uniquely determined by their respective syntactical structures[23]. Thus without a criterion to decide whether a given occurrence of the definite article is significant or not, we will end up like the proverbial Polish émigré logician who is reported to have remarked during his hastily arranged first formal lecture soon after his disembarkation at the Kennedy airport that in America there were "the tables" and "the chairs" but back home in Poland there were only simple but solid tables and chairs. Now following Langford (1) let us call the significant occurrences of the definite article its institutional uses; correspondingly we may call the non-institutional uses idiosyncratic uses.

At this stage we may drag in the reference to Meinong in Russell's communication to Quine which was quoted earlier. What sort of fallacies are those which Russell noticed in Meinong's work[24] and was drawing Quine's attention with the intention of avoiding them! Meinong held and defended the view that all occurrences of the definite article are institutional occurrences. Because of the multiplicity of definite articles in German, which is the vehicle philosophy, this in fact amounts to saying that all occurrences of all definite articles are institutional. Worse is that on this point Meinong did not differentiate between German and other languages; thus his thesis comes to mean all occurrences of all definite articles in any given language are institutional. This indeed was the belief of a philosophical tradition to which Meinong belonged and not invented; it existed before him, and it is now flourishing in the antipodes as well as on the other side of the Atlantic[25]. Russell with his robust sense of reality had a philosophical worry about all this; what, then, he wondered, can one do to account for the difference between terms like say 'the Sate' and 'a State'! If the two expressions mean the same then the distinction between the definite and the indefinite article breaks down, and if the distinction is to be preserved then we may have to, as Meinong in fact did, accommodate entities which can be denoted to by all terms with definite articles as their prenexes, such as 'the golden mountain', 'the husband of a spinster', and 'the class which is not a class'. More importantly, what does accommodating such entities amount to even if they are not full-blooded physical ones like bricks and bats, and are only conceptual, whatever that might mean and whichever way those conceptual entities are related or left unrelated to the tangible physical entities! Among other things, it amounts to subscribing to the view that rules of formation of expressions are at the same time principles of individuation of entities. To say the same in a simpler way, it amounts to holding that grammar generates the world, and that syntax produces semantics. In fact it can be shown to be implying worse, namely that each sentence produces its own truth-conditions. This unconstrained multiplication of entities merges with unrestrained idealism which is a variant of Hegelianism as well as Berkeleanism, perhaps. This implies that Meinongianism cannot be mobilized in defense of realism, as Russell during his brief flirtation with it though it could be[26]. In less than a year after he introduced Meinong to the English speaking world in 1904, Russell came to realize that his views thoroughly clinched realism. Thus his feverish search for a criterion to distinguish between the institutional occurrences of the definite article from its idiosyncratic occurrences.

In order to arrive at such a criterion, and to show that the criterion arrived at does meet the intended demands on it, the starting point obviously is the way in which the definite article occurs in communicative contexts and a characterization of its role in those contexts. Of course, the structure of a language has theoretical priority over the way it functions, but in constructing a theory of a language, its actual functioning may be allowed to take precedence; this is only granting investigative precedence which may be likened to moving from alibi to identify the criminal and, having identified the criminal, explaining how he could have left the alibi. More importantly, to preclude what needs to be isolated and till now has been eluding being opaquely reproduced in the projected characterization, the characterization itself will have to be carried out in a language which is different from the language to which belong the expressions that we propose to handle. Further it needs to be ascertained that the language in which the proposed characterization is to be carried out has adequate expressive power of the language the role of whose expressions is being characterized in it. That Russell was engaged for some years before and after 1905 in formulating such a language will not be news to anyone[27]; that this language as it was formulated in Russell (3) has adequate expressive power follows from the fact that all known mathematical theories were found to be expressible within the framework of that canonical language inducing faith in its adequacy to represent any indicative sentence of any other language[28].

Given this much we may begin, as Russell did, to observe the way in which the definite article figures in the indicative sentences of English to notice that it does occur in different ways, each of which is permitted by the grammar of that language. For example, it figures as a prenex expression of a subject term as in

(a) The African lion is ferocious

and

(b) The author of Waverley is Scott,

and as a prenex expression of the predicate term as in

(c) Scott is the author of Waverley,

and

(d) They are among the affected.

It even figures as a prenex of both the subject and predicate terms, for instance as in

(e) The tribals are the most exploited.

Further, as it can be noted from (a) - (e), it occurs as prenex of plural subject terms as in (a) and (e), and as a prenex of the singular subject terms as in (b). In fact one can find instances of its occurring as a prenex of a proper name as in

(f) The Ganges is terribly polluted.

There is no need to multiply examples or exhibiting the variety in the uses of the definite article in English[29]. So we can proceed to ask: which of these uses are institutional and which of these are idiosyncratic? A translation of these English sentences into a language whose vocabulary does not include the definite article or its surrogates will of course strain out all of its idiosyncratic occurrences in these sentences, and at the same time will indicate the contexts in which it occurs institutionally. Though a translation thus will allow us to isolate institutional occurrences of the definite article, it may not be sufficient to show whether in all of its institutional occurrences it behaves precisely in the same way or whether those occurrences are functionally different. For instance, translating (a) - (f) say into Hindi will reveal that in (f) the definite article occurs idiosyncratically but not whether it occurs in other cases, that is in (a) - (e), behaving in the same way. The reason for this is that the structure of the language from which a translation is effected, and the structure of the language into which it is translated both may be equally opaque requiring first clarifications, analyses and unpackings of the behaviour of the definite article, and secondly reductions and constructions in the language which is being used to present the translation. Now instead of unpacking each sentence in which the definite article occurs - obviously there are denumerably infinite number of them - we may as well think of a general principle, or a set of general principles of unpacking; such a principle would enable us to achieve what we intend to do through analysis by cases. Further, we may think of the principle with maximal generality such that it covers not only translations from a given language to another chosen language - say from English to Hindi - but from any language to any other language in so far as the sentences to be translated are in the indicative mood. After all, a thrust towards generality is intrinsic to thinking as such; and philosophers hardly need to be convinced that this is one of their cherished professional demands. Further, in order to be able to arrive at such a general principle, what needs to be done is to translate and represent different kinds of sentences of the English language containing institutional occurrences of the definite article into some canonical language whose structure, in virtue of its being finitely generated, is transparent. Because of this transparency the different ways in which the definite article occurs in different sentences and the relationship between those ways comes to be exhibited with clarity.

Opting for the canonical language of the Principia we can easily see that when the definite article occurs as a prenex of a plural subject term, it behaves like the Universal quantifier. Thus (a) above is the same as

(a.i) All African lion are ferocious.

Here 'the' is behaving not only like 'all' but exactly like it; this is to say that the scope of 'the' in (a) and the scope of 'all' in (a.i) are the same, namely the remaining part of those sentences, that is to say that 'African lions are ferocious'. Thus the canonical image of (a) is neither $(x)Lx \rightarrow Fy$ nor $(y)\{(x)Lx \rightarrow Fy\}$, but $(x)(Lx \rightarrow Fx)$. But this does not seem to be the case when 'the' figures as a prefix of a singular term which is the subject of a sentence[30]. In its institutional occurrence as a prenex of a singular term it is invariably followed by a predicate or a compound of predicates which constitutes its matrix; and, as we noted earlier, when its matrix is not a predicate but a proper name as in (f) it has neither an institutional role, nor a logical role. The definite article in this case cannot be said to be behaving like a quantifier because it does not satisfy the criterion that a quantifier in a quantificational sentence does normally satisfy. A quantificational sentence, say $(Q)\phi$- where (Q) is either (x) or (Ex) - is such that its truth-value is determined by the unique relationship that obtains between the quantifier and the matrix that figure in it; the instances of ϕ have a truth-value, and the truth-value of $(Q)\phi$ is uniquely determined by the truth-values of the instances of ϕ. In the case of a singular term with an institutional 'the' as its prenex, the matrix which is either a predicate or a compound of predicates is an expression to which truth or falsity cannot be meaningfully predicated; thus talking about the truth-value of the expression consisting of 'the' as its prenex and a predicate as its matrix being uniquely determined by an expression to which truth or falsity cannot be predicated does not make sense. Thus the definite article in its occurrences like this cannot be represented in the canonical language by the universal quantifier; the universal quantifier as it occurs in the canonical language is not a surrogate for the kind of the occurrences of the definite article in English that is being talked about here. Whereas 'the' in (a), as the universal quantifier, has the entire sentence as its scope, 'the' in (b) has no role to play in the predicate part of the sentence; it envelops only the subject part. Of course it affects the subject part, and in virtue of this has a logical role, but not in the way in which a quantifier has a role binding what falls under its scope. Its role here and in similar cases is exhausted by producing a singular subject term by saturating the predicate consisting of the matrix. Thus it is an operator, operating on the predicate (or the compound of predicates) occurring in the matrix. Before we move ahead to explore the logical nature of this operator, we need to preempt some possible confusions regarding what is involved in the operations which 'the' performs in such contexts. Such confusions are likely to crop when too much is read into these contexts, which is what more often than not philosophers are likely to indulge in under spells of extraneous associations - something which even Russell could not eschew.

That a singular subject term can be produced or constructed by an operation on predicate expressions does not bestow a privileged position on them. Nevertheless it can be used as evidence - be it plausible, or even infallible - in support of an anticidently and independently held belief of a Platonic sort to the effect that there were in the beginning properties or Forms. But what I have said in the foregoing does not imply a permission for predicates to usurp such a status. This is all the more so as singular nomial expressions too can be, in their turn, allowed to produce predicate expressions, and this should not induce the belief that in the beginning there were things. In fact, Quine's way of eliminating singular nomial expressions incorporates the very belief[1]. On the other hand we can afford to deprive predicate expressions all the privileges which they traditionally enjoyed. This shows itself in the manner in which sentences like (c) above are treated, or for that matter the way in which Russell treated them; these in fact are the same. There are, in English, two styles of "copulation"; 'is' thus indicates either predication or identity. This forces us to differentiate predicatively indicative sentences and identically indicative sentences. (c) above is an identically indicative sentence[32]. Thus there are no predicatively indicative sentences in which predicate terms have definite articles as their prenexes. To put it differently, predicate terms of predicatively indicative sentences do not contain saturated predicates. This means that an exhaustive explication of the logical role of the definite article can be carried out without any reference to operators operating on predicate terms, least of all quantification over predicates. The problem of definite descriptions, thus, is exclusively a problem in what we now a days call first order languages. As this pithy remark is likely to spark off two confusions, I hasten to offer two brief clarifications.

My setting aside higher order languages in understanding the logical behaviour of definite descriptions is informed by the fact that the problem at hand arises within the scope of first order languages, and that an invocation of higher order languages is of no avail. Lindström's results show why such an invocation is unrewarding. He proved in his (1) two theorems of great significance to philosophers of language; and these two theorems taken together establish that first order languages, in a well-definable sense of the term, are maximal. According to the first theorem there are no other languages apart from the first order one such that compactness theorem and Löwenheim-Skolem theorem both hold for them; and according to the second theorem, if a first order system is extended to result in say a system S with greater expressive power, and if compactness theorem and Löwenheim-Skolem theorem both hold for it, then S will not be having an adequate proof-procedure, this is to say that it will be devoid of an adequate set of rules of deduction. There are two aims of a logical system; they are first to facilitate representation of statements and arguments - specially mathematical ones for they constitute a test case for the adequacy of the system, and secondly to provide a powerful tool for investigation - again, mathematical investigation because it provides the experimentum crucis for judging the adequacy if the system. Second order systems may be of some help to mathematics with respect to the first aim, but not the second[33]. Try to imagine mathematics without a rigorous proof-procedure! I am inclined to go with Wittgenstein to hold that what is unique to and in mathematics is its method of proof[34]. So I refrain from invoking second order languages. But this may cast a shadow on my banking so much on the canonical language of the Principia; so a word in defense of that choice.

The canonical language of the Principia is multi-ordered, obviously in virtue of its incorporating type-distinctions. So the legitimacy of translating sentences from ordinary language into the language of the Principia may be suspected, I anticipate. It might be thought that my decision to opt for this language, and my intention to keep second order languages out of court are rather incompatible. But there should be no worry on that count for two specific reasons. First, in the language of the Principia quantification is restricted to individuals. Evidently, because of type-distinctions, the realm of individuals is not homogeneous, and hence the problem of saturated subject terms arises at each level of the ordered universe of individuals. In any case, we know how to evade type-ambiguities, and there is the axiom of reducibility to help us in pushing the problem downwards[35]. It is true that this is a controversial axiom; but we need not concern ourselves with it at least here[36] because all that is required for the present is a recognition of the fact that the canonical language in question can be set free from the theory of types and the related controversial strategies by reformulating it as one whose adequacy for representing at least all known mathematical theories has been well-established for long[37]. Thus we can afford to translate sentences of ordinary language into the language of the Principia and understand the logical structure of sentences containing definite descriptions. In the next chapter I shall take note of the way in which Russell himself handled such a translation.

Notes

[1] Russell virtually convicts Strawson on that account. In Russell (11) he points out how out of his pet pair of sentences 'The present King of France is bald' and 'Scott is the author of Waverley', Strawson concentrates on the former to find fault with his analysis of it, and how he ignores the latter as it does not suit him.

[2] See Russell (2); in his (1) Strawson responded to Russell (2) and also to (3) and (4). Though he referred to all of them together, he cited and quoted from Russell (3). It is surprising that without exception editors of anthologies juxtapose Strawson (1) and Russell (2), or occasionally Russell (4), but never Russell (3).

[3] Of course, this integration is based on the assumption that theories of singular terms, proper names and definite descriptions, are independent of quantification theory, or that nomial links of languages with the world fall outside the scope of logic whose exclusive concern is with syntax. Differing with the Russell-Wittgenstein realistic view to the contrary, van Fraassen holds that syntax is ontologically neutral and hence the questions of validity cannot be decided on the basis of syntactical form. There are other difficulties with van Fraassen's approach. See Chapter Five below, and Rao (5).

[4] To note further implications of this drawback see Bencivenga (1); van Fraassen, however, is quite aware of these limitations of his approach. See f.n.16 to his (1).

[5] Why Strawson did not think of reformulating or replacing the received language is beyond my comprehension; nor do I understand how Strawson would respond to the Wittgensteinian view that the vague, even the vague, can be given precise margins. After all, there are rigorous theories of chaos.

[6] It may be the case that ordinary language is such that incompatible theories of it are equi-possible, and thus Aristotelian, Russellian, and van Fraassenian theories are equally applicable to it. It may even be the case that, on the strength of Strawsonian reasoning, it is not amenable to logical analysis and adjudication beyond a certain point or level. But whether or not ordinary language eludes logical analysis is itself a logical question open to logical investigation. The answer to this question cannot be prejudged by ordinary language itself, as Strawson believes it to be and thereby attempts to close the issue. In refusing to allow logical investigation he is being a misologist and is betraying his contempt for reason.

[7] In contrast, ontological realism amounts to the thesis that there are sentences which are not members of any such set, or variantly that there are sentences which cannot be formulated in any language whatsoever. For instance, consider the sentential schema 'x is a transcendental' and let x range over reals P_x, such that for $i < j$, P_i and P_j are substituends for x in the schema. Thus we have sentences, true or false, 'P_i is a transcendental', 'P_j is a transcendental'. Now let R be the set of such sentences. R obviously is a nondenumerable set, but the set of sentences S of a language is a denumerable one. Hence some members of R are not members of S. This is to say that there are sentences which cannot be formulated in any language. We may call them Finsler sentences after the Swiss mathematician who first thought of them in the early twenties; see van Heijenoort (2). The existence of Finsler

sentences can be established by using induction and elementary laws holding for enumerable sets. Let $L_1,\ldots,L_k,L_{k+1},\ldots$ be an enumeration of languages, and let S_i be the set of sentences of L_i. For $i < j$, if S_i and S_j are recursively enumerable, so is $S_i \cup S_j$. Let US^k be the union of all sets up to and including S_k. By induction it follows that the result holds for US_k. It does not matter whether $L_1,\ldots,L_k,L_{k+1},\ldots$ itself is nondenumerable. If it is, we can use the axiom of choice to generate the sequence on which complete induction can be applied to prove the intended result.

Dummett could have taken the bull by his horns by attacking van Fraassen on this point. Instead he chose a metadisciplinary point of the latter to rebut his antirealism. See 'van Fraassen Against Scientific Realism' in Dummett (1).

[8] Of course the idea as well as the explanations and justifications can be traced to Wittgenstein who wondered if it is neither true or false, what else a sentence could, after all, be! It can be found even in Aristotle; see Rao (1). I mention Dummett as he is the most recent well-known defender of the idea.

[9] I do not think that Strawson is aware of the implications of van Fraassen's work for in our brief personal talks on the issue he created the impression that he was happy at the fact that van Fraassen provided a sophisticated logical explication of, and offered the much needed support to, his notion of presupposition. But this is in the nature of things; when one's enemy appropriates his weapons he hardly notices it until they are used against him.

[10] In fact Russell himself challenged his critics by demanding them to prove the relevance of all that they have published in philosophical periodicals towards understanding "the $\sqrt{-1}$ is $\sqrt{-u}_{/2}$"; see Russell (1).

[11] I am using this term in its literal sense and am contrasting it with 'natural language'; so it should not be taken with all of its Quinean affiliations.

[12] Let L be a language in some non-canonical notation; let ψa and $\psi(Ix)\phi$ be sentences of L where a is a proper name and $(Ix)\phi$ a definite description. Within the language of the Principia the corresponding fungible sentences respectively will be: $(Ex)(\psi x . x = a)$, and $(Ex)\{\psi x . (y)(\phi y \equiv x = y)\}$.

The canonical notation as conceived by Quine however does not permit such a straightforward construction of fungible sentences. It requires first to extend L augmenting its vocabulary by singular predicates and deleting proper names to result in L*, and then construct fungible sentences of ψa and $\psi(Ix)\phi$ in L via L*, as $(Ex)(\psi x . P'x)$, where P' is the singular predicate introduced corresponding to the deleted proper name a, and $(Ex)\{\psi x . (y)(\phi y \equiv y = x)\}$.

The syntactical difference between a and $(Ix)\phi$ in L is reflected even in L*; for instance the predicates occurring in the corresponding fungible sentences here are from L* only, whereas the predicate terms that occur in the first pair of sentences are from the vocabulary of L. Thus the first pair of sentences is in $L \cup L*$, whereas the second pair is in $L \cap L*$.

[13] For other kinds of attributive uses of proper names see Boer (1).

[14] Emphasis in the original; the idea that concepts constitute the denotata which is explicit in this passage may be ignored for the present.

[15] I hope that it was not for this reason that Kripke was maximally polemical in his (2) while dismissing Tharp's justification of substitutional interpretation of the quantifiers. Further, dragging in modal notions and modal logic to solve problems concerning definite descriptions is either to complicate the problem or to evade it altogether. Modal logic and possible world semantics survive by shifting problems - by sacrificing hypothetical sentences about the actual in favour of categorical sentences about the possible. This should induce us to be skeptical about attempts to extrapolate results from more technical work in modal logic to handle problems with which we are concerned here. Kripke however cannot be accused of doing it; otherwise his (1) would have been entirely different. But this is done by Kripke aficionados. Concerning their contributions it may be said that in virtue of the fact that Kripke modal logic, with Kripke semantics, does not include open sentences as theorems. Because of this their utility in solving problems of singular terms is nil.

[16] Goldfarb (1), p.354 commenting on this says: "One must not be misled by Russell's verbiage surrounding the introduction of his formal apparatus. His prose cannot count as producing an explanation, much less a 'semantics', for the system, because it involves violations of type restrictions. Hence it is nonsense, according to the basic principles of the theory of types. The point was emphasized by Wittgenstein and Ramsey. Frege becomes involved in analogous difficulties, illustrated by the notorious problem that the concept horse is not a concept. ... He says, ruefully, "by a kind of necessity of language, my explanations, taken literally, sometimes miss my thought ... I fully realize that in such cases I was relying upon a reader who would be ready to meet me half way - who does not begrudge a pinch of salt." In what follows I shall be discussing Russell's theory meeting him half way and not begrudging.

[17] See Rao (5); I take this to be the fate of all those who while presenting their discoveries resort to narrating the process of their discovering them. In such writings what was discovered gets mixed up with how it was discovered. The desirable way of presenting philosophical discoveries thus seems to be the way of Wittgenstein in the Tractatus; the only alternative is the method of Carnap. Is this why Plato advised philosophers to be wary in talking to hoi pollai !

[18] 'PM' here can be an abbreviation only of 'Principia Mathematica', and not of 'Principles of Mathematics'; Russell started his work on definite descriptions which culminated in his (3) only after the Principles has been published. His treatment of definite descriptions in the work is Meinongian in intention and pseudo-Fregean in execution. 'The A' in the framework of the Principles is semantically autonomous as much as a proper name is. It has, in fact, 'meaning' in isolation; and its meaning is a "denoting concept." Russell accepted what is called Frege's Constitution Principle, according to which the meaning of a sentence is a function of the meanings of the semantically autonomous terms that figure in it. To account for the semantic autonomy of terms Russell, following Meinong, postulated concepts. It is only with his theory of descriptions that Russell came to save his robust sense of reality by overcoming the need to postulate concepts, and the need to preserve the semantic autonomy of definite descriptions. More of this in the next chapter.

[19] What could have been, I wonder, Russell's response had Strawson focused his attention only on Russell (2).

[20] Thus neither Lambert (8) nor Consuegra (4) is taking Russell's contribution in its entirety. Lambert's view that Russell had two theories of descriptions, one in Russell (2) and another in Russell (3), that the former is concerned with ordinary language whereas the latter deals with formal language, and that according to the former definite descriptions are singular terms which go as subject terms of indicative sentences whereas in the latter they are "incomplete symbols" cannot be textually supported; Consuegra is right on this point. But he is far from being fair to Russell in holding that he was preoccupied with the same issues in both (2) and (3). Russell's point is that the logical problem barely raised in (2) but solved in (3) is so to say the dog, and the bunch of philosophical problems handled in (2) is the tail. Consuegra has shown how on the basis of Russell's hitherto unpublished work it emerges that Russell had on his mind the notion of an incomplete symbol since 1905. In any case, though the expression 'incomplete symbol' does not figure in Russell (2), it is all pervasive in it that definite descriptions are semantically opaque and stand in need of an analysis. Further, the analysis he proposed for them in (2) banishes them from the realm of subject terms. This goes against Lambert's view.

[21] I have not qualified the last name by the epithet 'early', as Wittgenstein subscribed to this view even three years before his death. I may cite a passage of 1948 vintage from Wittgenstein (6): "Philosophy is a logical affair; yes, it is comforting to realize that philosophical confusion signifies that I do not know my way about in logic; for there are only important problems and not petty ones." Of course, it can be pointed out that there is some equivocation in the use of 'logic' here; but I trust that it does not affect the issue at hand. Wittgenstein may have used 'logic' in this context in a wider sense than he used it in the Tractatus, though in view of his (2), part I, sec.97 this can be doubted. Anyhow, the anticipated objection will be too weak to call into question my broadly correct remark; so I let it pass without further discussion.

[22] See Russell (9). Citing a 1947 work in an explanation of a problem Russell though that he solved in 1905 might appear anachronistic. But this is not so for this criticism of Hegel antedates his (2). Russell continued to subscribe to this view even after the war. Moreover, I am not interested in history of ideas or in the evolution of a research program. My intention here is to note the conceptual relation between ideas, arguments, explanations, and theories. As such I can afford to do what I have done, and it will be objected to only by those who are either bewitched by history, or else are ignorant of it.

[23] This may be taken as my explanation of the much misunderstood Wittgensteinian remark that any language is good enough and that no language needs our apology.

[24] Russell is referring to Meinong (1) which he reviewed in his (1).

[25] Routley at Canberra, Lambert and Parsons on the west and the east coast of America respectively are the leading philosophers belonging to this tradition. See Routley (1), Lambert (4), and Parsons (1).

[26] See Russell (1).

[27] It would be fruitful to see Russell's theory of descriptions as a part of such an effort, an important part for that matter because this program initiated in (8) got into troubles, some of which are overcome with the help of the results arrived at in (2); this smoothened the path to (3).

[28] Developing this canonical language has not been an aprioristic activity; it was as much inductive as is the development of any other scientific theory. The syntax and semantics of this canonical language were justified alike by bringing them to mutual agreement. Its syntax was amended if it resulted in something which went against cultivated semantic intuitions, and its received semantics was altered if it was found to violate the accepted syntax. The development first order language from Frege (1) to Hilbert and Ackermann's formalisation of it in their (1) through Peano and Russell and many others show just that.

[29] Those interested in this may consult Robbins (1) and Yotsukura (1).

[30] It is heartening to note that even Strawson is convinced that "the logic of ordinary speech provides a field of intellectual study unsurpassed in richness, complexity, and power to absorb;" Strawson (2), p.232. But he refuses to admit that this is precisely why margins of precision need to be gained for it. In fact he subscribes to the gloomy view that such a logic can never be systematically described.

[31] Church's lamda operator produces singular terms from sentential schemes, and Leonard's dot operator produces general terms from sentential schemes; see Suppes (1), and Leonard (2). No metaphysical or ontological theses should be read into any of these syntactical formulations. Nevertheless, it must be admitted that each of them will come out as a perfect fit for some metaphysical thesis, or ontological doctrine.

[32] It is well-known how Russell attacks Hegel for confusing predicative occurrences of the copula with the copula of identity. For example, he writes on p.42 of his (7): "Owing to this confusion he (Hegel) thinks that 'Socrates' and 'mortal' must be identical. Seeing that they are different, he does not infer, as others would, that there is a mistake somewhere, but that they exhibit 'identity in difference' ... This is an example of how for want of care at the start vast and imposing systems of philosophy are built upon stupid and trivial confusions, ..."

[33] My argument earlier that there are sentences which cannot be represented in any language may be better appreciated now, as it should now be clear that what I meant was that there are sentences that cannot be represented in any first order language whatsoever. See Barwaise's contribution to his own (1).

[34] See Rao (5).

[35] In all the millions of words printed about the theory of descriptions there is not even one that shows an awareness of the fact that Russell's theory rests on the axiom of reducibility. It is not for nothing that he developed his canonical language devoting the twelfth step to this

axiom, the thirteenth to identity, and the fourteenth to definite descriptions. These are nested. In that language identity will not be available without the axiom of reducibility, and without identity definite descriptions cannot be defined. See Rao (5). Wittgenstein attacked this axiom, and rejected the identity theory of the Principia; this naturally led him to reject Russell's theory of description. More of these matters will emerge in the following two chapters.

[36] See Rao (5).

[37] Though we can easily isolate the first order part of Russell's canonical language, Russell himself does not do it. This perhaps is due to the fact that he, at least until he came into contact with Wittgenstein, took logic to be about something, namely everything. The laws of logic have content and are truths about the logical structure or furniture of the world. For instance, on p.169 of his (4) he writes: "Logic is concerned with real world just as truly as Zoology, though with more abstract and general features." Here Russell is following the trend set by Frege who in 1882 declared that his "intention in (his (1) of 1879) was not to represent an abstract logic of formulas, but to express content through written signs in a more precise and clear way than is possible to do through words."

Chapter Two : Russell's Riddles

To isolate what Russell came to notice as the basic logical problem concerning descriptive functions and to delink it from what he thought to be extraneous philosophical issues, I shall take into account his (8), (2) and (3) - specially, his introduction to the second edition, and chapters 12, 13 and 14 - and also (4) and (12); I shall ignore as far as possible and as much as it is necessary all secondary literature with moribund quibblings over what Russell meant by 'mean', whether 'denoting' and 'meaning' are the same or different, or whether 'denotation' and 'reference' can be used indifferently, and similar issues. Towards the close of his discussion of the definite article in (4) Russell writes that

> " ... for the purpose of mathematics, the more philosophical parts of the theory are not essential."[1]

As Russell did not differentiate between logic and mathematics either at the time of developing his theory of descriptions or at the time of writing this, it follows that, as he thought, his discussion of philosophical issues is not essential either for formulating the logical problem or for solving it. Further, he seems to have been convinced that the problem as it arises in the realm of logic is conceptually prior to the way it shoots in other areas, for he remarks in (2) that it is of

> "very great importance not only in logic and mathematics but also in theory of knowledge."

And having announced his theory, paradoxically first to philosophers and in a professional journal of philosophy he implores the reader

> "not to make up his mind against the view - as he might be tempted to do, on account of its excessive complication - until he has attempted to construct a theory of his own on the subject. ... This ... attempt ... will convince him ... that whatever the true theory may be, it cannot have such a simplicity as one might have expected beforehand."

But what is the logico-mathematical problem and how did Russell come to notice it? What after all was the conceptual necessity that motivated the invention or the discovery of the theory of descriptions? As sufficient attention does not seem to have been paid to these issues, and as by now what Russell wrote concerning these issues has either been forgotten or else simply ignored - of course, at the cost of a fair understanding of the issue as well as Russell - devoting some effort to these problems seems to be in order.

It was an integral part of a long tradition beginning with the Promethean myth[2] and culminating in Kronecker's belief, moving through the Pythagorean metaphysique, that the notion of an integer is indefinable. Weierstrass initiated a movement in which this tenet was rejected; the theory of cardinal integers envisioned by him, and developed by Dedekind and Cantor, finally resulted in Frege's and Peano's definitions of the notion of an integer. The Promethean tradition requires us to use integer terms as subject terms; we may call this a <u>mathematical object theory</u>. Having shown in his (2) why such a theory was unacceptable, Frege developed what can be called a <u>mathematical property theory</u> according to which integer terms occur in the predicate part of a sentence, strictly speaking in the function part of a sentence. He took arithmetic as a part of logic, and showed how arithmetical sentences can be defined in logical terms. In his (2) he gave a brief sketch of such a reduction of arithmetic to logic. Though Russell differed with Frege as to the way in which he developed the logic required for the proposed reduction, and in fact showed how Frege's way of doing it results in paradoxes in virtue of its being too strong, he subscribed to the basic tenets of Frege[3]. Peano[4], on the other hand, having rejected the received view that the notion of an integer was indefinable, attempted to define it in terms of Dedekind's postulates which we now refer to as Peano axioms. His intention however was not to reduce arithmetic to logic. What he did was to deduce arithmetic from a set of suitably chosen axioms characterizing the system of integers. In this sense Frege's theory was superior to Peano's, for it enables us to prove the existence of the system of integers instead of simply assuming it. More importantly, Peano's theory was still a mathematical object theory; what it shows is that, given an integer and some characteristic, or generic, properties of integers, the whole system of integers can be generated; thus the rest of the members of the system of integers can be defined in this narrow sense of the term. In this restricted sense Peano's theory is at the same time a mathematical property theory. The postulates are tagged to a suitable system of logic to facilitate deduction. Thus arithmetic turns out to be an axiomatised extension of logic. But Peano had a very narrow conception of logic; relations, according to him, fall outside its scope, and belong to specific theories. Because of this he had to include in the set of axioms he proposed for formulating arithmetic what we now consider as identity axioms along with Dedekind postulates[5]. Peano's postulational definition of the system of integers, as Russell calls it, consists in his showing that the system in question satisfies those axioms. But there is a rub, and Peano himself was aware of it. The postulates are not satisfied by the system of integers uniquely. This is to say that the system, as it has been formulated by Peano and is familiar to us, is not the only system which satisfies those axioms. Peano himself noted that

> "(t)here is an infinity of systems satisfying all the primitive propositions. All the systems which satisfy the primitive propositions have one-one correspondence with numbers. Number is what is obtained from all

> these systems by abstraction, in other words, number is the system which has all the properties enunciated in the primitive propositions and those only."[6]

This puzzled Russell on two counts; he wondered as to

> "(i) How are the various systems which agree in satisfying the propositions be distinguished?"

and whether there was

> "(ii) any process of abstraction from all systems satisfying the five axioms ... logically possible?"[7]

Peano however thought that such issues concerning his postulational definition can be handled by definition by abstraction. But after a careful scrutiny Russell found that Peano was wrong and there was no way of circumventing the issues by invoking definitions by abstraction. Peano's argument was that each of those infinite number of number systems satisfying his postulates could be one-one related to any of those systems, and that this could be done without requiring the notion of unity, this is to say that without indulging in circularity. The strategy suggested by him towards that end was: take number as a property of classes, and take two classes having the same number as their being in one-one relation; then let two classes be similar when they have the same number. This relation of similarity[8] is reflexive, symmetrical and transitive; and these properties, according to Peano, indicate that when this relation holds between two terms, the two terms so related have a common property, and that if they have a common property then the relation of similarity holds between them. This common property is their number. In short, this is Peano's definition of number by abstraction, as Russell understood it. But applying the definition by abstraction over the infinite number of systems which satisfy Peano's postulates does not give us the system of integers with which we are familiar, as Wittgenstein would have put it, "our system of integers." All that it may yield is that these infinite number of systems are similar. Thus Peano's attempts have not resulted in a definition of our system of integers; they have succeeded in showing that this system is a member of a similarity class. Nevertheless this amounts to characterizing the notion of an integer as:

> The system of integers with which we are familiar is a member of a similarity class, each member of which satisfies Peano's postulates.

But this does not constitute a definition; it falls short of being a definition. One of the important reasons for which it fails to be a definition is that what may be identified as the definiendum contains the definite article, and what may be identified as the definiens contains an indefinite article; Frege commented on such purported definitions in his (2) with which Russell is familiar. While saying the same, but not so pointedly, Russell refers to Peano's work on definability - rather indefinability, as we shall note soon - and thinks that the passage in question fails to be a definition as the uniqueness of the presumed definiendum is not carried over to the presumed definiens. The copula in the passage is predicative, and does not stand for identity. He thus holds that Peano's

> "definition by abstraction, and generally the process employed in such definitions, suffer from an absolutely fatal formal defect; it does not show that only one object satisfies the definition."[9]

To be able to arrive at a definition which carries the uniqueness embedded in the definiendum to the definiens, it is necessary to show that the similarity in question is not only an analytical property of the class of systems satisfying Peano's postulates, but also is an analytical property of each member of that class. This means that it requires to be shown that each inductively generated subsystem of each of these systems is a member of a similarity class. No two members of the latter similarity classes - let us call them secondary similarity classes - are similar. Each of them is unique. Now, this dissimilarity, and consequently this uniqueness, is indefinable within the framework of Peano's definitional strategy. As such his

> "definition wholly fails to define the number of a class."[10]

To sip the mouthful of coffee beneath the whole lot of formal foam, Russell's remedy was to define the number of a class as the class of all classes similar to the given class, and consider membership of this class of classes as a predicate in order to show that it is satisfied by all and only similar classes. Equipped with this remedial measure, Peano's definition can be made to work in the intended direction. Taking our integer system as a point of reference - as the given class, to use Russell's term - and constructing similar classes for each inductively generated subsystems of it, we will be able to define a number as a property of a class. But this is not possible within the framework of Peano, for it does not, as Russell did notice

> "admit relations into the fundamental logical apparatus of logic."

Thus neither that the integer system is the class of all systems similar to our system of integers, nor that an integer is the class of all classes which are similar to the given class having a unique relation to that integer and nothing else, can be brought out from Peano's formal apparatus.

There is yet another difficulty with Peano's framework. He held that each mathematical theory has its own indefinables, and that the definables of a theory are defined in terms of the indefinables of that theory and their properties. Thus for instance, though 2 can be defined as the successor of 1, and 1 as the successor of 0, zero itself remains as the indefinable of his system. After all, the entities satisfied by the basic step of an inductive definition will have to be assumed as given. For Russell, on the other hand, there are no mathematical indefinables; the mathematical property theory was almost an act of faith with him. The indefinables for him are exclusively logical. However this issue should not be of much concern for us here; so we shall pay to it only marginal attention. Our intention here is not to highlight Russell's contribution towards a definition of the notion of a number; moreover there is not much that I can add to the existing literature on that. In any case his logicist attempts have been unsuccessful, and were proved to be unsuccessful. My primary interest here thus is in noting his contribution towards understanding the behaviour of the definite article. Yet, his work towards a definition of the notion of a number culminating in his unsustained thesis that all mathematical objects are "logical constructions" needs to be taken into cognizance as it motivated him to work on the theory of descriptions. We may then note in brief how this came about.

From Russell's remark that Peano's definition by abstraction is not "a logically valid process," and that it has serious "formal defects," it follows that he had definite views on what constitutes a formally perfect and logically impeccable definition. He had a conception of a definition, though he did not invent it; he borrowed it from Padoa[11] to whom Peano himself owes much and whom Russell met at the Paris Congress of 1901 along with Peano. In fact he gives in his (8) a Padoan definition of a mathematical (logical) definition, and tried to set it apart from philosophers' definition[12]. Whereas what is considered as a definition in philosophy, according to Russell, is an analysis of an idea - complex or composite, obviously - into its constituents, the logico-mathematical definition does not propose any such analysis; further, the definiendum of a logico-mathematical definition need not be a composite expression defying initial understanding but could as well be a simple and perfectly intelligible one. Moreover the composition and the content of the definiendum of a logico-mathematical definition are irrelevant for the definition being logically valid or otherwise because it concerns itself only with the <u>form</u> of the definiendum. As Russell characterizes logico-mathematical definitions, what sets them apart from other definitions even in the vulgar sense of the term is that they show how

> "given any set of (logico-mathematical) notions, a (logic-mathematical) term is definable by means of these notions, when and only when, it is the only term having to certain of these notions a certain relation which itself is one of the said notions."[13]

Having recognized this, Russell declared his decision:

> "I shall ... in future ignore the philosophical sense and speak only of mathematical definability."[14]

He seems to have stuck to his decision in the entire series of writings concerning the problem at hand; and we need to keep this in mind when we consider his attempts to define definite descriptions in (8), (2), (3) and (4).

This induced Russell to offer in (8) what he thought was a logically valid definition of the notion of a number; he thought that this definition was incremental over the one given by Peano. Soon we will note how what he thought was wrong and how what he held was not legitimate; but before that we need to take into account that having envisioned a scheme to reduce the entire realm of mathematical objects to logical notions, Russell seems to have hit upon the idea that by deploying this strategy of logico-mathematical definition to another area he would be able to redeem himself from the burden of the Meinongian monsters. It may be recalled that Meinongianism looms large in his work of that period, specially in (8). The fourth chapter of this work is outright Meinongian; it subscribes to the view that every syntactically well-formed expression is semantically autonomous. But this will lead to anomalies specially in the case of expressions like 'the even prime other than 2'; after all, 'the', 'even', 'prime', 'other than', and '2' are all in the primitive vocabulary of the language in which the theory of integers is formulated (by a working mathematician mixing some colloquial with formalism), as much as 'Golden' and 'mountain' are in the primitive vocabulary of the ordinary language, and the syntactical rules of the language of that theory do permit the formation of the expression in question. Why not then, instead of resorting to

Meinongianism to account for these expressions, evade the anomalies by defining away such expressions, and define them away in the logico-mathematical sense of the term! This is to say that the apparent semantic autonomy of such expressions can be reduced to their real semantic dependence on the unproblematic autonomy of some other expressions by resorting to the strategy of offering them logical definitions. Why not pick up a set of logical notions and a logical relation which does not include the logical notion which figures in expressions of this sort, namely 'the', and then show how it is related to those notions by the relation in question? Thus Russell found an antidote to Meinongianism by logically defining 'the' in terms of 'some', 'all', 'and', 'if-then', and 'is identical with'. The outcome was his (2) incorporating an initial and informal sketch of his theory of definite descriptions. As the program of reducing all mathematical notions to logical ones which he envisioned was too vast a scheme requiring considerable time and effort, and as it was easy to sketch the byproduct, namely the strategy for getting rid of Meinongianism, he attended to it first and announced his theory first to philosophers, and through the same medium that he used to express his admiration for Meinong, namely Mind; this created the wrong impression that his philosophical investigations influenced his invention of the theory of descriptions. Anyhow, the discoveries announced in his (2) set Russell's thought in motion. He felt the need to show that identity was a logical relation, and that in turn required him to go for the axiom of reducibility. Thus the final statement of the theory of descriptions in Chapter 14 of Russell (3) is preceded by an argument towards the legitimacy of the axiom of reducibility in Chapter 12, and the theory of identity which presupposes the axiom in Chapter 13.

Before considering this theory in detail, it is desirable to delink it from the historical and personal background in which it was thought of; it needs to be isolated from the logicist program of reducing mathematics to logic. What Russell succeeded in establishing is not that the notion of an integer can be defined exclusively in terms of logical notions, but that in a theory which includes his theory of types, the axiom of reducibility, and his theory of (numerical) identity, the notion of a number can be defined in the sense that a mathematical property theory, whose underlying logic is the quantification theory as developed in the Principia[15], can be constructed such that each analogue of a provable sentence of the mathematical object theory of Peano can be proved in it. (As far as the unprovable sentences are concerned, both the theories are subject to Godel's limiting result.) To give a balance sheet of his effort, he succeeded in showing that

1. Theory of types + axiom of reducibility + theory of identity + theory of descriptions . ⇒ . theory of integers,

and

2. theory of types + axiom of reducibility . ⇒ . identity theory + theory of descriptions.

He did not succeed in showing that

3. quantification theory . ⇒ . theory of types + axiom of reducibility + identity theory + theory of descriptions

There is no reason to believe that he showed that either

4. quantification theory . ⇒ . theory of types + axiom of reducibility,

or that

5. quantification theory . ⇒ . identity theory + theory of descriptions.

This should throw some light on why Russell had to put in the formidable effort that he did put in to defend the logicality of the theory of types and the axiom of reducibility, and to extend the boundaries of logic beyond the limits drawn by Peano, to include in it these and the theory of identity, thus in fact whatever is required for making a property theory of integers feasible. Thus Russell showed what he wanted to show by assuming all that he needs to show! As I intend to return to the axiom of reducibility and the theory of identity when I take into account Russell's final formulation of the theory of descriptions in his (3), I note here only that the logicality of the theory of descriptions has these credentials[16].

It is well-known how having defined a number as "the class of all classes which are similar to a given class," Russell tried to overcome the assumption that classes exist - an assumption which is essential to Peano arithmetic[17]. He did it by developing what he called a no-class theory, this is to say that, besides other things, by defining away classes - strictly speaking class expressions - in terms of quantifiers, sentential functions, identity, truth-functional connectives, and the diadic relation of membership $\in$, where $a \in y$ is a well-formed expression when a is an individual term, and $y = \hat{x}\phi!x$, i.e. the set of all those terms such that a substitution of each of which for the variable in the sentential function $\phi!x$ yields a true sentence. We need not recall this story here, let alone assess its merits and demerits, for our interest here is only in Russell's banishing descriptions and not exiling classes. For this purpose it is sufficient to note that in working out this program Russell discovered two important features about definite descriptions.

The first among the two features of definite descriptions which Russell noticed is that "in all propositions about 'the so and so' there is an apparent variable."[18] Russell's distinction between an apparent variable and a real variable may be brought about as follows in purely syntactical terms[19]: when vacuous quantification is not permitted, to say that ϕ contains a quantifier is to say that ϕ is of the form $(Qx)\psi$, where Q is either a universal or an existential quantifier, and x occurs free in ψ; and the occurrence of x in the quantifier is apparent, whereas its occurrence in ψ is real. Now to say that a sentence in which 'the class of all classes which are similar to a given class' is the grammatical subject contains a quantifier is the same as to say that this definite description contains a variable, and that the variable figuring in it is bound by the quantifier in question. The second feature of definite descriptions which Russell seems to have noticed is that in so far as a real variable is present in them, they are like functions. If there is a unique entity answering that definite description, then there is a unique value of the corresponding function with the variable x as its sole argument. Russell was familiar with the work of Boole and Schorder who showed how the variable x can be eliminated from sentences that can be written as equations like $F(x) = 0$ in terms of polynomials like $ax + b\bar{x}$ where a and b are logical expressions in which x does figure[20]. Russell was not only familiar with the work of these logicians, but was aware of its limitations as well. That seems to be the reason for his turning to Peano for inspiration and direction. Concerning his indebtedness to Peano, Russell wrote:

> "His great merit consists not so much in his definite logical discoveries nor in details of his notation (excellent as both are), as in the fact he first showed how symbolic logic was to be freed from undue obsession with forms of ordinary algebra and thereby made it a suitable instrument for research".[21]

So the variable occurring in a definite description and the quantifier occurring in the sentence containing that definite description as its subject term will have to be eliminated by definition. Nevertheless, Russell's insights which influenced his handling of definite descriptions and (non-denoting) proper names, specially his understanding of their logico-linguistic behaviour, have their sources in algebra. Consider for instance a constant operation $f(x)$, where f is a short for x 0, multiplication by zero; for all arguments of $f(x)$, the value of this constant operation[22] is the same, namely 0. Thus $f(x) = 0$; as such the constant 0 can always be replaced by $f(x)$. With reference to such constant operators it can be said with certainty that the following invariably hold:

(i) $(x)(Ey)\{f(x) = y\}$, and

(ii) $(x)(y)(z)[\{f(x) = y \,.\, f(x) = z\} \rightarrow (y = z)]$.

Russell seems to have noted the similarity between the behaviour of such constant operators and the behaviour of definite descriptions on the one hand, and the similarity between the behaviour of constants and proper names on the other. The following passage from his (2) induces us to entertain that possibility.

> " ... (I)f 'c' is a denoting phrase it may happen that there is one entity x (there cannot be more than one) for which the proposition 'x is identical with c' is true ... We may then say that the entity x is the denotation of the phrase 'c'. Thus Scott is the denotation of 'the author of Waverley'."

We may now ask whether a constant operation can be replaced by an individual constant; the answer, obviously, is that it can be done provided the constant operation in question satisfies a condition, namely that it is the case that for each of the arguments of the variable figuring in the function, the value of the operation is the constant in question. This is to say that only when conditions corresponding to (i) and (ii) above are satisfied, $f(x)$ can be replaced by 0. This means that though a name can be eliminated by a definite description without any difficulty (if need be, taking 'being called so and so' as a predicate![23]) a definite description cannot be replaced by a proper name unless it is a proper description, this is to say that it is one which satisfies the two conditions in question. This implies that proper names and definite descriptions are different kinds of expressions - and idea which dominates Russell's writings from (2) to (3). The reasons for this being so are in fact embedded in the conditions which a definite description will have to satisfy so as to be able to perform, in the context of a sentence, in effect the role of a proper name. It succeeds in performing this role only when its denotation is unique and when its denotatum exists; we may call these respectively the uniqueness condition and the existence condition. A definite description may fail on either of these counts, or even on both these counts. In the case of an expression which looks like a proper name but turns out to be falling short of meeting the demands of uniqueness and existence, it will have to be concluded that it really is a truncated definite description. There are no improper names in the strict sense of the term, though there are improper definite descriptions; some of these are evidently so, like 'the even prime different from 2' for instance, and others are opaquely so, like 'Pegasus' for example. As such that two proper names are identical, or that they have the same denotatum, if true is trivially so, but that two definite descriptions are identical, if true is significantly so; and so is the case with a sentence in which the identity sign is infixed between a proper name and a definite description. Further, what has been noted till now about definite descriptions holds good with grammatical proper names like 'Socrates' and 'Pegasus'. There is a quantifier and a variable implicit in sentences in which proper names like these occur as subject terms; for instance 'Socrates is a philosopher', on analysis, thus becomes - this is the term Russell uses in his (2) - (Ex)(x = Socrates, and x is a philosopher). So Russell declared in his (2):

"I take the notion of a variable as fundamental."

The fundamental indicators, as Russell intended to say, are variables - bound variables as Quine[25] subsequently made explicit and precise what is vague and implicit in Russell. However, we shall not consider proper names separately because first the problems concerning definite descriptions themselves are sufficiently taxing, and secondly if a solution to these problems is gained, then it can be deployed to solve the problem concerning proper names (either by constructing a unique monadic predicate corresponding to each proper name[26], or by showing the equivalence of a bound variable with a unique value and a proper name.)

Though syntax guides us in deciding whether an expression is a definite description or not, it is insufficient to tell us whether a definite description is proper or improper. Identifying an expression as an indicative one is a syntactical affair because, as Russell would say

> "a phrase is denoting solely in virtue of its form."[27]

In the case of definite descriptions which are of our primary concern, we can easily secure a purely syntactical criterion to tell whether a given expression of that sort is an indicative one or not, for instance the following: all expressions beginning with an occurrence of the definite article followed by a predicative function are indicative; thus all expressions of the form 'the x such that ψx' come to be indicative. We may now start using the letter I in capitals to represent the logical use of 'the' , and say that all expressions whose form reduces to $(Ix)\phi x$ are indicative expressions. Russell's point concerning expressions of the form $(Ix)\phi x$ is that they are not semantically autonomous, this is to say that they do not indicate on their own and that their indicative role is derivative of their being logically related to the fundamental indicators (bound variables, that is) occurring in sentences in which those expressions figure as subject terms. This means that the variables figuring in definite descriptions are not fundamental indicators. The reasons for which Russell does not grant semantic autonomy to expressions of this form are too well-known to be recounted here in detail. I may offer some clarifications about them because they are as well-known as they are misunderstood and even misrepresented.

Russell explicitly states in his (3) that

> "whenever the grammatical subject of a proposition can be supposed not to exist without rendering the proposition meaningless, it is plain that the grammatical subject is not a name directly representing some object."[28]

Now, $\psi(Ix)\phi x$ is of the same form as ψa, and it does not become "meaningless" - this is to say that it will not be devoid of sense - if $(Ix)\phi x$ is not indicative. As sentences of the same form cannot be said to be about different realms of objects (specially one about senses, and the other about denotata - to use Frege's terminology), and as all indicative sentences are about actual objects indicated by the fundamental indicative terms occurring in them, it must be the case that the real subject terms of sentences containing expressions of the form $(\imath x)\phi x$ as their subject terms are to be elicited and exhibited after analyzing them[29]. Moreover granting semantic autonomy to expressions of the form $(\imath x)\phi x$ results either in Meinongianism or in Fregeanism. The former leads to logical errors (contradictions), and the latter to artificial devices (anomaly averters)[30]. Thus definite descriptions will have to be taken as semantically dependent[31]. But though no expression of the form $(\imath x)\phi x$ has "any significance of its own," it no doubt "is essentially part of a sentence"[32] in which it occurs. Thus though $(\imath x)\phi x$ is not semantically autonomous, it is a part of, a significant part of, say some $\psi(\imath x)\phi x$; this means that in the context of its occurring as a part of $\psi(\imath x)\phi x$, $(\imath x)\phi x$ comes to have a semantic role. If this is the case, what it indicates in that context will have to be elicited by fixing what the sentence $\psi(\imath x)\phi x$ is about; and this can be known only when $\psi(\imath x)\phi x$ in its entirety is analyzed into quantificational language such that the fundamental indicators occurring in the resultant analysans may reveal what the sentence is about. Thus Russell came to see the importance of offering contextual definitions of definite descriptions. Evidently contextual definitions need to be explicit definitions; and Russell was already convinced of the fruitlessness of implicit definitions, and how they fall short of the intended goals. Peano's definition of the notion of an integer in terms of a set of postulates is a case in point. But there are difficulties, as Russell noted in (2), with contextual definitions of definite descriptions. These definitions will be structurally different from similar definitions of say quantifiers or predicates, and will have a distinct and differentiating feature of their own. For instance, the contextual definition of a quantifier, say (x), involves only an apparent variable and truth-functional connectives as in say

$$(x) \ldots \equiv_{df} \sim(Ex) \sim \ldots .$$

This is a contextual definition in the sense that the definiendum figures as a part of the definitum, this is to say the left hand side of the definition. The definitum in a contextual definition of a predicate contains variables, and will be contextual in the sense that it constitutes a syntactically well-formed expression in which the predicate in question, which is the definiendum, occurs as a part. But in a contextual definition of an operator the constant figuring in the definitum contains the sign of identity too. A contextual definition of a definite description is not intended to define the predicate which is saturated by the operator that the definite article is; on the other hand it is geared to define the variable figuring in it taking a unique value. In defining 'the x which authored Waverley" we are not defining authoring Waverley, but only x as it satisfies

(a) $(x)(Ey)(Ay = x)$,

and

(b) $(y)(z)(x)[\{(Ax = y).(Ax = z)\} \rightarrow (y = z)]$

It may be noted that Ax here is not figuring as a sentential function but only as a saturated predicate expression. This sort of an occurrence of an expression may be likened to the following: consider for instance

(i) Plato is bald,

and

(ii) The philosopher whose most famous student is Plato is bald.

'Plato is bald' occurs in (i) as well as in (ii), but in different senses of the term; in (i) it is semantically autonomous, and in (ii) it is not so - it is not even a semantic part of (ii)[33]. It goes into the formation of a syntactically well-formed unit which is semantically autonomous, and thus can be said to be a part of that unit. In short, it is a compositional part. Similarly, in $\psi(\mathtt{Ix})\phi x$, $(\mathtt{Ix})\phi x$ is a compositional part, and only the former is a semantically autonomous whole. $(\mathtt{Ix})\phi x$ as it occurs in $\psi(\mathtt{Ix})\phi x$ cannot be taken to be indicating something as Meinong took it to be, nor can it be taken to be expressing a sense and hence having an indication as Frege took it. If it is understood in this way, besides facing the "curious difficulties" which the relationship between the "sense" and the "indication" of an expression would lead to, we will be forced to go in for artificial devices in the case of those definite descriptions which are devoid of indication, for example 'the even prime which is different from 2'. Further, because the class of indicative sentences will have to be syntactically determined on the basis of their form alone $\psi(\mathtt{Ix})\phi x$ and $\psi(\mathtt{Ix})\chi x$ both will have to be provided indication even if one of them say $(\mathtt{Ix})\chi x$ is improper and $(\mathtt{Ix})\phi x$ is proper. Otherwise the principle of parity of form cannot be held. Thus when one of them is improper, we are forced, as Frege was, to institute an artificial indication for it and similar expressions, for instance by choosing the null set to serve the intended purpose. The other way is to "abandon the view" - which subsequently came to be called Fregean Constitution Principle - that the indication of a sentence containing a definite description is not a function of the indication of the description. Thus the difficulties which we will have to face when we go for options offered by Meinong and Frege constitute, according to Russell, only evidence for the legitimacy of his choice. They do not comprise the reasons for which he chose what he did choose. What Russell adduces as the major reason for his choice is that a phrase of the form $(\mathtt{Ix})\phi x$

> '<u>per se</u> has no meaning <u>because</u> in any proposition in which it occurs the proposition, fully expressed, does not contain the phrase".[34]

The phrase itself can be "broken up" by fully expressing a sentence of the form $\psi(\mathtt{Ix})\phi x$.

Thus Russell's theory of descriptions amounts to a procedure of reducing all sentences of the form $\psi(\mathtt{I}x)\phi x$ "to forms in which no (such) phrases" occur[35]. As all sentences of the form $\psi(\mathtt{I}x)\phi x$ are not of the same complexity, the projected reduction cannot be uniform and straightforward, which is why Russell thought that the reduction is "interesting and difficult." This necessitated an understanding of first what is involved in sentences of the form $\psi(\mathtt{I}x)\phi x$ solely in virtue of their having the form they have, and secondly which constituents embedded in form not containing phrases of the form $(\mathtt{I}x)\phi x$ take care of what is involved in sentences of the form $\psi(\mathtt{I}x)\phi x$. The implicit claim of $\psi(\mathtt{I}x)\phi x$ is that there is a value of x which instantiates ϕ, and nothing else instantiates ϕ, or variantly, if anything else instantiates ϕ, then it is numerically the same as the value whose existence we have already affirmed, and that the value of the variable x instantiating ϕ whose existence has been affirmed instantiates ψ also. Thus definite descriptions essentially involve numerical identity; more importantly, when $(\mathtt{I}x)\phi x$ occurs as a term (or as a grammatical subject) in a sentence, its import is not just that

(1) $(Ex)(y)\{(\phi y \equiv (x = y)\}$;

on the other hand it is that

(2) $(Ex)\{\psi x \,.\, (y)(\phi y \equiv x = y)\}$.

It is evident that (1) and (2) are not in general equivalent, for (1) may be true and (2) may be false if ϕ contains another definite description. Intending to return to such complex cases a little later, let us for the present allow ϕ to be atomic. When ϕ is atomic and is instantiated by a value, (1) and (2) come out as equivalents. Now to say that an atomic predicate ϕ is instantiated is the same as to say that the value of the variable x in ϕ exists. Thus taking 'exists' ($E!$, for short) as a predicate of one argument, and taking it so only for argument sake, we can say that $E!(Ix)\phi x$ and $(Ex)(y)\{(\phi y \equiv (x = y)\}$ are equivalent[36]. Further, in so far as there are no uninstantiated atomic predicates in the language of the Principia, and in so far as E! is not a primitive predicate of that language[37], when ψ is a primitive predicate in it, $\psi(Ix)\phi x$ "becomes" $(Ex)[\psi y.(x)\{\phi x \equiv (x = y)\}]$. This is to say that it reduces to a sentence which contains only primitive (instantiated) predicates and variables, and does not contain the operator I. Thus the reduction is unambiguous when the given sentence contains only one definite description. This however is not the case when more than one definite descriptions occur in a sentence.

Consider a simple case, for example $\psi(Ix)\phi^1x(Ix)\phi^2x$, where ψ is a predicate of two arguments. Unless it be the identity predicate, the two orders in which the two definite descriptions can be eliminated issue in two different sentences which are not logically equivalent. Thus the sentence in question is ambiguous. The situation gets doubly complicated when ϕ^1 and ϕ^2 themselves contain definite descriptions; and more serious ambiguities come to the fore if ψ is not atomic. For instance let $\psi(Ix)\phi^1x$ be C ψ^1 $(Ix)\phi^1x$ where C is a monadic sentential connective - be it logical or epistemological. As our aim is to understand Russell's puzzles, we need not consider epistemic connectives; so let C be the tilde standing for negation. Then C $\psi^1(Ix)\phi^1x$ may become either

(I) $(Ey)[\sim\psi^1y.(x)(\phi^1x \equiv x = y)]$,

or

(II) $\sim(Ey)[\psi^1y.(x)(\phi^1x \equiv x = y)]$.

(I) and (II) are not logically equivalent unless it be the case that $E!(Ix)\phi x$. Given $E!(Ix)\phi x$ we can prove the equivalence of (I) and (II). This is to say that

$$\{E!(Ix)\phi x\} \rightarrow \{(I) \equiv (II)\}.$$

But to assume E!(Ix)ϕx, and then reduce ψ^1(Ix)ϕ^1x into canonical language is not carrying reduction solely on the basis of its form, for in doing so we will be going beyond the bounds of syntax and will be resorting to a semantic, if not an empirical, assumption. And if the projected reduction is to serve the intended goal, each given sentence needs to be unambiguously reduced to a unique sentence of the canonical language of the Principia. So a strategy needs to be designed to disambiguate reductions; this is to say that the order in which the different definite descriptions occurring in a sentence are to be eliminated, and the scope of each of the definite descriptions occurring in it needs to be syntactically fixed - fixed, that is, entirely on the basis of the from of the composition of sentences.

There has been a lot of dissatisfaction concerning the strategy Russell designed for overcoming the hurdles in connection with the problem of scope[38]. But taking into account Russell's basic insight, his strategy can be reformulated and an inductive definition can be offered towards a specification of the order in which definite descriptions can be eliminated from a given sentence. Deferring the reformulation to a later stage in the enquiry, we shall note here some basic features of Russell's strategy, even if it, on the first glance, appears rather clumsy. Russell pinned down the reasons for the ambiguity of the sort we mentioned above with respect to the syntax of the sentence in which definite descriptions occur. As he states it in his (3):

> "when (Ix)ϕx occurs in a proposition which is a (syntactical) part of a larger proposition, there is doubt whether the smaller or the larger proposition is to be taken as the ψ{(Ix)ϕx}."[39]

Treating ψ(Ix)ϕx as the scope of (Ix)ϕx, Russell suggests that the scope of a description be indicated by prefixing an apparent occurrence of the definite description to the sentence coinciding with its scope, and then using punctuation marks suitably to preclude possible confusions. Thus

$\{(\text{Ix})\phi^1x\}\sim\psi^1(\text{Ix})\phi^1x$ is the sentence which on reduction "becomes" (I) above, and $\sim\{(\text{Ix})\phi^1x\}\ \psi'(\text{Ix})\phi^1x$ is the sentence which reduces to (II)[40]. The scope operator thus indicates the extent to which the reduction is to be carried out at a given stage; and in case a sentence contains multiple occurrences of definite descriptions, the initial ambiguity can be overcome by following the natural convention which assigns all terms minimum scope, thus allowing the definite description which occurs first larger scope, and the definite description which occurs last the smallest scope[41]. Thus if ψ is atomic, there is no ambiguity in $\psi(\text{Ix})\phi x$; it must then be $\{(\text{Ix})\phi x\}\ \psi(\text{Ix})\phi x$. If ψ is complex then the scope of $\{(\text{Ix})\phi x\}$ is the whole of $\psi(\text{Ix})\phi x$. In the case of $\psi\{(\text{Ix})\phi^1x\ (\text{Ix})\phi^2x\}$ - "unless the contrary is expressly indicated," as Russell would say[42] - it is evident that it must be

$$\{(\text{Ix})\phi^1x\}\ [\{(\text{Ix})\phi^2x\}\{\psi(\text{Ix})\phi^1x\ (\text{Ix})\phi^2x\}]$$

Thus the notion of scope can be extended on the lines indicated to cover cases in which connectives with more than one place and quantifiers are present leading to more ambiguities. And if $\sim\psi(\text{Ix})\phi x$ is to be the negation of $\psi(\text{Ix})\phi x$ then it will have to be $\sim\{(\text{Ix})\phi x\}\psi(\text{Ix})\phi x$, because $\{(\text{Ix})\phi x\}\sim\psi(\text{Ix})\phi x$ is not the negation of $\{(\text{Ix})\phi x\}\psi(\text{Ix})\phi x$. In this respect the scope operator behaves like the universal quantifier, which too in a sense is a scope indicator; compare the way in which $\{(\text{Ix})\ \ldots\}$ and (x) behave in this regard. The scope operator and the negation operator are related in the same way as the universal quantifier and the negation operator are related[43]. Obviously, $\sim(x)\phi$ is not the negation of $(x)\sim\phi$. It is for this reason saying that the iota operator is a term-forming operator - this is to say that it is something which when prefixed to an open sentence results in a term - is not correct. As it emerged by now, it is much more that that, and has well-defined relationships with other quantifiers; for instance $(x)\phi x \rightarrow \{(\text{Ix})\phi x\}\ \phi(\text{Ix})\phi x$ is as much a logical truth as is $(x)\phi x \rightarrow (Ex)\phi x$. If the description in question is proper, and if improper descriptions are permitted to be used

$$[(x)\phi x \, . \, (Ex)\{x = (Ix)\phi x\}] \rightarrow \{(Ix)\phi x\} \; \phi(Ix)\phi x,$$

and

$$[(x)\phi x \, . \, (Ex)\{x = (Ix)\phi x\}] \rightarrow (Ex)[\{x = (Ix)\phi x\} \, . \, \phi x]$$

come out as logical truths. (Of course, as has been hinted at earlier, under certain conditions definite descriptions behave like terms. Given $(x)(Ex)(\phi y \equiv x = y)$, where x is free in ϕ, scope indicators can be dispensed with. $\sim(Ix)\phi x$ may become or be a short for $\{(Ix)\phi x\}\sim(Ix)\phi x$ or $\sim\{(Ix)\phi x\}(Ix)\phi x$. Then $(x)(Ex)(\phi y \equiv x = y) \rightarrow [\{(Ix)\phi x\}\sim(Ix)\phi x \equiv \sim\{(Ix)\phi x\}(Ix)\phi x]$ can be deduced as a theorem of the Principia.)

The reduction proposed by Russell is to be carried out on those sentences in which the scope operator unambiguously indicates the part of the sentence consisting its scope. Thus $\{(Ix)\phi x\}\psi(Ix)\phi x$ becomes $(Ex)\{\psi x \, . \, (y)(\phi y \equiv x = y)\}$. In so far as this reduction is carried out independent of the content of the definitum, and the composition of either the definiendum or the context in which it occurs, solely on the basis of the form of the definitum, it can be turned into a schematic definition, that is a definition of a sentence of the form figuring as the definitum. Thus

$$[(Ix)\phi x]\psi(Ix)\phi x =_{df} (Ex)[\psi x \, . \, (y)(\phi y \equiv x = y)].$$

Even now the stage is not set in all the required details for contextually defining the phrases of the form $(Ix)\phi x$; there still are some formidable problems which first need to be understood and then solved. The logical import of sentences in which definite descriptions occur, for instance $\psi(Ix)\phi x$, is that they involve uniqueness. This is evident to anyone who knows English, though Russell who I trust knows it pretty well is said to have borrowed it from Frege and Peano, who the learned historians of logic say discovered it. Anyhow, what this amounts to is that each of the sentences containing definite descriptions is satisfied by at least one and at most one, thus exactly one object. Thus, $\psi(\mathrm{Ix})\phi x$ cannot be true unless both (i) $(\mathrm{Ex})\psi x$, and (ii) $(x)(y)\{(\psi x . \psi y) \rightarrow (x = y)\}$ are true.

> "These two together are equivalent to $\psi x \equiv_x x = y$ which we may define as $E!(\mathrm{Ix})\phi x$. Thus "$E!(\mathrm{Ix})\phi x$" must be a part of what is affirmed by any proposition about $(\mathrm{Ix})\phi x$."[44]

This clearly shows that $\psi(\mathrm{Ix})\phi x$ essentially involves identity; and one can raise several issues concerning Russell's projected contextual definition of definite descriptions on this count. It may be recalled that he wanted to contextually define them such that the definiens in any such definition consists of, besides all the predicates contained in the definitum, only quantifiers, truth-functional connectives, and identity. It may be pointed out that in so far as identity figures both in the definiendum and the definiens, the definition may fall short of being in perfect order, all the more so because while claiming how his definition of an integer is incremental over that of Peano, Russell emphasized the fact that logico-mathematical definitions should contain in the definiens a relation not contained in the definitum. Moreover, what is the nature of identity itself! Is it a relation? Or is it a function? Is it logical or extra-logical? Russell did think a lot about these issues, and his writings on these issues are confusing and conflicting; but what he finally came to do with identity in (3) is crystal clear. For instance in his introduction to the first edition of (3), having mentioned that it has the three usual properties of reflexivity, symmetry, and transitivity, Russell writes that

> "We shall find no new definition of the sign of equality is required in mathematics."

This poses more problems than those that Russell thought that it solves. If no new definition is required, why did he propose one? Or is it that what he offered was not a definition in his sense of the term? Is Russell, on the other hand, conflating identity with equality? Obviously √4 and 2 are equal; but are they identical! it is tempting to raise this issue as Russell requires the identicals to be inter-substitutable, and hence they should be substitutable in sentences like '2 is an integer' which is true to arrive at a sentence like '√4 is an integer' which is false on a specific interpretation. But then one can explain away these problems by pointing at the fact that Russell's aim was to define the notion of an integer, and thus in the language in which the proposed definiens figures there will be no sentences in which 2 will figure as a subject term, and hence '2 is an integer' is not a sentence of that language at all.

That apart, what is identity? Frege did raise two important issues relating to it, namely whether it is a relation, and if it is a relation whether it is a relation between terms or objects indicated by them![45] Russell felt the need to settle these issue, and the need to show that identity is a logical notion which is required to reduce arithmetic to logic. Thus his puzzles about identity from (8) to (3).

> "The question whether identity is or is not a relation", Russell wrote[46] echoing Frege, "and even whether there is such a concept at all, is not easy to answer. For it may be said, identity cannot be a relation, since were it is truly asserted, we have only one term, whereas two terms are required for a relation. And indeed identity, an objector may urge, cannot be anything at all: two terms plainly are not identical, and one term cannot be, for what it is identical with? Nevertheless identity must be something. We might attempt identity from terms[47] to relations, and say that two terms are identical in some respect[48] when they have a given relation to a given term. But then we shall have to hold either that there is strict identity[49] between the two cases having identity in the sense of having a given relation to a given term; but the latter view leads to an endless process of an illegitimate kind. Thus identity must be admitted, and the difficulty as to the two terms of a relation must be met by a sheer denial that two different terms are necessary. There must

> always be a referent and a relation, but these need not be distinct; and where identity is affirmed, they are not so.
>
> But the question arises: why is it even worthwhile to affirm identity? If we say "Edward VII is the King" we assert an identity; the reason why this assertion is worth making is that in one case the actual term occurs, while in the other a denoting concept[50] takes place … Often two denoting concepts occur, and the term itself is not mentioned, as in the proposition 'the present Pope is the last survivor of his generation'.[51] When a term is given, the assertion of identity with itself, though true, is perfectly futile, and is never made outside the logic books; but where denoting concepts are introduced, identity is at once seen to be significant. In this case, of course, there is involved, though not asserted a relation of the denoting concept to the term, or of the two denoting concepts to each other. But the is which occurs in such propositions does not itself state this further relation but states pure identity."

This passage is no paradigm of transparency in Russell's philosophical writings; yet, it is clear from it that for an identity sentence to be meaningful it is not necessary that both the terms linked in it need not be terms. One of these can be a denoting concept, a notion which Russell abandoned soon after (8) because of its Meinongian affiliations, and replaced it by the notion of a denoting phrase, which finally was changed to the notion of an incomplete symbol in (3). It may even be the case that in a given sentence both the expressions related by identity are incomplete symbols, and the real term which is indicative may not be mentioned at all. The identity - the "pure identity" - between the two incomplete symbols is determined by the relation of these symbols to the real term which is left unmentioned. If this term indicates a unique object on the one hand, and on the other is uniquely related to both the incomplete symbols, then the given identity sentence is true; and if either of them fail on the uniqueness condition, then it is false. Thus a sentence in which the copula in singular is infixed between two definite descriptions, first is an identity sentence, and secondly if it is true, it is true not because both the definite descriptions are proper and indicate the same object, nor because both of them happen to have a given relation to a term, but because there happens to be one and only one object which is indicated by the term to which the two given definite descriptions have a unique relationship. Given that a sentence in which a definite description occurs requires the existence of such an entity, it follows that the truth of an identity sentence containing two definite descriptions connected by the copula in singular requires that the unique entity whose existence is required for the truth of a sentence in which one of the definite descriptions occurs be numerically the same as the unique entity whose existence is required for the truth of a sentence in which the other definite description occurs. This is to say that the truth of $(\mathrm{I}x)\phi x = (\mathrm{I}y)\psi y$ requires the truth of $x = y$ where x and y are not the variables figuring in the identity sentence in the example but are among the variables figuring in its reduction sentence.

Here $x = y$ is still a "numerical sentence" as = here is expressing numerical identity; and the very purpose of the program as a part of which the theory of descriptions was thought out was to reduce numerical truths to logical ones. To say that x and y indicate one and the same entity is to still use the conceptual battery of the ancien régime. Having reduced integers to classes, and having defined away classes predicatively[52], numerical truth itself stands in need of a definition in terms of predicates, Further, unless such a definition is offered, it cannot be said that arithmetic has been reduced to logic. This propelled Russell to define numerical identity in terms of an exclusively logical apparatus, this is to say that in terms of quantifiers, individual variables, predicates, and truth-functional connectives. Why Russell needed identity was precisely why any of us needs it, namely to have substitutivity conditions for terms. He noted, as was stated earlier, that substitutivity follows from the three properties of identity, that is from its reflexivity, symmetry, and transitivity. So he set himself the task of characterizing substitutivity in terms of the logical apparatus he devised, and define identity - numerical identity that is - in terms of substitutivity. Though Russell's design is clear - it is well indicated in the example he gives, namely the identity of 2 and √4 - there is a lot of confusion in what he says about the problem involved in defining identity; at least there is a clear swapping, conscious or otherwise, of numerical identity and identity in other senses of the term of which we shall say enough soon.

> "It is plain", he writes, "that if x and y are identical, and ϕ_x is true, then ϕ_y is true. Here it cannot matter what sort of function $\phi\hat{x}$ may be; the statement must hold for any function."[53]

But this results in an anomaly; as $2 = \sqrt{4}$ is true in the numerical sense of =, and as '2 is an integer' is true, '√4 is an integer' must be true. Of course, in view of what Russell says about the converse of what is affirmed in the passage quoted above, the anomaly here can be overcome. So let us ignore it and move to consider whether substitutivity implies numerical identity. Can it legitimately be said that x and y are identical only if <u>for all</u> ϕ, $\phi x \rightarrow \phi y$! The quantification involved in this is "inadmissible", for it leads to a vicious circle and a contradiction. This is what the theory of types is all about.

The theory of types does not permit a definition of identity in terms of substitutivity for it involves the illegitimate unrestricted quantification over predicates. A way out is to restrict quantification over predicates, and quantify over predicates of one specific order consistent with its type. $x = y \equiv \phi x \rightarrow \phi y$ holds good for any ϕ of the given order.

> "Thus we obtain, so to speak, a hierarchy of different degrees of identity",

as Russell would say. But identity is an unambiguous notion; the identicals are identical, and the different are different. There are no degrees of identity as Hegel and his followers whom Russell was fighting would have us believe. Russell has an argument to show how this obtains. Suppose that for each n, we have $x = y \equiv \phi x \rightarrow \phi y$ where ϕ contains predicates of order n. Even then we cannot legitimately say that x and y have all predicates in common and hence are numerically identical because that involves quantification over predicates which is not restricted to an order. But it would be a fair bet to believe that if $x = y \equiv \phi x \to \phi y$ is true in the case of all predicates of order n, then it is true in the case of all predicates of the lowest order. For $n = 1$, it is obviously true; and for $n = 2$, it is true because having predicates of the lowest order, this is to say of order 1, is itself one of the second order predicates. Now suitably indexing sentences to indicate clearly the order of predicates occurring in a material equivalence sentence we can say that

$$\{(x = y) \equiv_2 (\phi x \to \phi y)\} \to \{(x = y) \equiv_1 (\phi x \to \phi y)\}.$$

Using induction and the transitivity of material implication, we may arrive at

$$\{(x = y) \equiv_{n\geq 1} (\phi x \to \phi y)\} \to \{(x = y) \equiv_1 (\phi x \to \phi y)\}.$$

This means that the nth order substitutivity of say x and y implies their first order substitutivity. We have as yet nothing at our disposal on the strength of which we may be able to assert the converse of this, namely

$$[(x = y) \equiv_1 (\phi x \to \phi y)] \to [(x = y) \equiv_{n\geq 1} (\phi x \to \phi y)].$$

And unless we have that we cannot say that the identicals are universally substitutable; but in order to be able to have that, we need

$$(\phi x \to \phi y) \equiv (\phi x \to_{n\geq 1} \phi y)$$

which, on any count, is not a logical truth. In fact it is of the form $\phi \equiv (\phi = \psi)$, and is in effect the so-called reducibility postulate in one of its possible formulations[54].

But given all this, specially the reducibility postulate, identity can be defined in terms of substitutivity as follows:

$$(x = y) =_{df} (\phi)(\phi x \to \phi y)$$

with the only proviso that ϕ, x, and y satisfy the demands of the theory of types[55]. Now, what does the theory of types amount to? We may note that this is not all that is given to it; but this is all that we need to take into account in so far as it touches upon the issue at hand, namely the problem concerning substitution, identity, and definite descriptions. The theory of types gives rules to fix the permissible substitutions for say x in ϕx; to put it differently it gives a procedure to set apart the substituends of x in ϕx from all other expressions which cannot be legitimately substituted for x in ϕx. As Russell remarked, after gaining sufficient intelligibility about his own creation, it all comes to saying that

> "a word or symbol may form part of a significant proposition, and in this sense have meaning, without being able to be substituted for another word or symbol in the same or some other proposition without producing nonsense".[56]

Thus the theory of types essentially is geared towards a determination of the set of well-formed expressions which are syntactically categorisable as sentences. This in fact is what Russell himself came to recognize by 1918, perhaps under the influence of Wittgenstein. He wrote in his (10) that the

> "theory of types is really a theory of symbols, not of things. In a proper logical language it would be perfectly obvious. The trouble arises from our trying to name what cannot be named. If we had a proper language, we should not be tempted to do that."[57]

If our language were perfect, this is to say that if its syntax were fully expressed, the theory of types would be dispensable, because then it would be redundant[58]. The theory of descriptions requiring the theory of types, the axiom of reducibility and theory of identity based upon this axiom, was necessitated by the imperfections of our language, specially its syntax. What Russell thus thought that he was doing was just bringing a little more perfection to our language by paying attention to its syntax - its grammar; echoing Wittgenstein he remarked in his (10):

> "I think the importance of philosophical grammar is very much greater than it is generally thought to be."

Notes

[1] Emphasis added

[2] See Aeschylus' Prometheus,

> "I gave them number
> The chief stragem of all."

Kronecker is reported to have said that God created numbers, perhaps on the eighth day of creation!

[3] See Appendix to Russell (8).

[4] My knowledge of Peano is based on secondary sources of which Russell's work itself is one. Peano's 1889 Arithmetices Principia has been translated into English only in part by van Heijenoort in his (2); but his four volume 1889 - 1903 Formulaire de mathématiques is not as yet available in English. Peano was reported to be unaware of Frege's work, though Frege took note of Peano's 1889 work soon after its publication.

[5] This fact was used as evidence for holding that Peano lacked familiarity with Frege's work in logic, and that his conception of logic was influenced by the writings of Boole and Schorder. See Consuegra (3).

[6] As quoted by Russell in his (8); emphasis added.

[7] Russell (8)

[8] I am emphasizing the demonstrative pronoun so that the notion of similarity in the present context may not be confused with the philosophical, epistemological notion i.e., the notion of similarity which is weaker than identity, and is not transitive. The relation of similarity in question is that of equinumerosity.

[9] Russell (8); emphasis added.

[10] ibid.; emphasis added.

[11] Padoa's semantic method, which he used to prove the indefinability of multiplication in terms of addition and 1, and the indefinibality of addition in terms of the successor function and 1 was valued to be of great importance by Peano; its fecundity was exploited during the twenties by Tarski and Lindenbaum, and later by MaCinsey and Beth. See Beth (1), and De Pouverve (1). Those who want a brief account of this method may see Suppes (1).

[12] It is 'philosophers' definitions' and not 'philosophical definitions'; the latter would suggest that there are philosophical definitions as there are mathematical definitions, just as there are

African elephants and Indian elephants. Philosophers' definitions and mathematical definitions are not definitions of different species of the same genera. What philosophers pass

as definitions are not really definitions. They are explanatory devices with heuristic value. They may be called definitions by courtesy, at most.

[13] Russell (8); emphasis and interpolation ours.

[14] ibid.; p.112

[15] Russell calls this the theory of apparent variables; but we need not be priggish about the inessential.

[16] See Rao (5).

[17] In fact Peano's original formulation of the principle of induction quantifies over classes.

[18] Russell (3), Vol.1, p.50.

[19] There is a semantic distinction between apparent and real variables, and it is more complex. In this sense of the term the variable that figures in the sentential function, say the x in $\phi!x$, is a place-marker for a variable nomial expression which can be substituted for x, a for instance. The subsituend a is the real variable, and the x in the function is a place-marker for the real variable, thus an apparent variable.

[20] In fact Boole claimed and Schorder proved a stronger thesis according to which all sentences can be expressed as logical polynomials. It may be recalled that it was Boole-Schorder algebra that influenced Peano who in turn influenced Russell. Though Peano was unaware of Frege's logical writings, Russell was familiar with Frege (1), but worked out his logic under the influence of Peano and independent of Frege. See Sainsbury (1), Grattan-Guinness (1), (2), and (3), and Goldfarb (1).

[21] Russell (3), Preface; it is time for a similar redemption again; only this time it needs to be freed from abstract algebra and model theory.

[22] Constant operators should not be confused with constant functions like multiplication by 1; the value of a constant function for each argument is the same as the argument itself.

[23] Though this is distasteful to several philosophers, Russell relishes it.

[24] I wonder why Russell switched over to the term 'denotation' rather than stick to 'indication'. He used the pair 'indication' and 'meaning' to translate Frege's "Bedeutung' and 'Sinn'; see Appendix to his (8). It could be that Meinong's writings were fresh in his memory.

[25] See Rao (5).

[26] Russell's 'is called Socrates', or Quine's 'Socratises' is such a predicate. Here the difference between proper names and definite descriptions come to the fore. That Scott is called 'Scott' is trivially true; but this is not so with either that the author of Waverley is called 'Scott' or that Scott is called the author of Waverley.

[27] Russell (2); emphasis in the original.

[28] Chapter III; emphasis added.

[29] Given this, whether Russell has been successful in "proving" that expressions of the form $(\mathtt{I}x)\phi x$ are meaningless, or whether he means by 'meaning' sense or denotation, seem to be besides the point. Yet there was a lot of debate on these issues. See, for instance, Coddy (1). Russell himself was partly responsible for this affair. He explained his point to different groups at different levels, to non-professional audience rather imprecisely. Coddy cites one such explanation from Russell (12) to show that he did not succeed in giving a "proof" to the effect that definite descriptions are "meaningless". But the very idea of a "proof" is irrelevant in this context, for what Russell was attempting to work out was a coherent system of semantics - I say this despite Goldfarb (1) - which he envisioned as an alternative to the systems offered by Meinong and Frege.

[30] See Russell (2).

[31] In his Introduction to the second edition of (3) Russell gives yet another reason for treating expressions of the form $(\mathtt{I}x)\phi x$ as incomplete symbols; it is that $(x)\psi x \rightarrow \psi(\mathtt{I}x)\phi x$ is not always true. In the language of the Principia $(\mathtt{I}x)\phi x$ is neither a primitive term, nor is it a sentence; thus it is an incomplete symbol as complete symbols are either primitive terms or sentences. It may be noted that this characterization of an incomplete symbol is completely syntactical. The complete symbol is: $\{(\mathtt{I}x)\phi x\} \ldots (\mathtt{I}x)\phi x \ldots\}$. Thus $(\mathtt{I}x)\phi x$ is at once a non-bindable singular term, and a self-binding singular term. $(\mathtt{I}x)\phi x$ is ill-formed, like $P \rightarrow Q$, perhaps; the untruncated and well-formed version of it would be $(P \rightarrow Q)$.

[32] ibid.; emphasis in the original.

[33] Though both of them contain 'Plato' and 'Plato is bald', only the first is about Plato and the second is about Socrates.

[34] Russell (2). The passage shows the basic difference between Russell and the ordinary language philosophers. According to Russell, sentences of ordinary language are perfectly in order in virtue of the received grammar; only neither that grammar is fully specified, nor is it free from imperfections. Thus sentences of ordinary language are grammatically opaque, for this grammar itself is so.

[35] Russell (2); emphasis added. 'Reduction' is the term that Russell uses.

[36] For a clarification of this notion of equivalence see Chapter Four below.

[37] To say that E! is not a primitive predicate of the language of Principia is the same as to say that it does not contain sentences of the form E!(Ix)ϕx; variantly put, (Ex){E!y.(x)(ϕx ≡ x = y)} does not figure in the Principia.

[38] p.173; interpolation ours.

[39] This prefixing the definite description to indicate scope and treating the whole of the definite description as an "apparent variable" apparently shows how a definite description behaves like a variable, this is to say that as an indicative term. Thus the variables figuring in the definite description should not be taken as fundamental indicating terms of the language into which the reduction is carried out. Wittgenstein thought that Russell did not note this basic factor and brought it to Russell's notice; but it should be evident from what we have shown, namely that Russell in fact based his analysis on this point. See Wittgenstein (3) and Rao (5).

[40] It should be evident that if all the descriptions occurring in a sentence are proper, it makes no difference which one is eliminated first and which one next. An ordering of eliminations thus is required as we do not know, prior to elimination, whether a definite description is proper or improper.

[41] See Chapter Four below for further discussion on this point and also on Hintikka's skepticism concerning the possibility of fixing the scope of a definite description.

[42] This is rarely done in ordinary language; it is left to be taken care of by the proficiency of the users. In fact the ambiguity in question arose because of the syntax, the grammar of that language, is not completely specified.

[43] Geach (2) remarks that according to Russell (3) there is no univocal negation; but this does not correctly represent Russell's view. The fact of the matter is that there is a univocal notion of negation which needs to be defined for unanalyzed and analyzed sentences separately. This does not mean that there are two kinds of negation embedded in the language of the Principia, propositional negation and predicational negation, nor does it imply that there are in it two kinds of sentences.

[44] Russell (3), second edition, Introduction, p.68.

[45] Rao (5).

[46] Russell (8).

[47] What does it mean to say that identity is a term! Of course, it was treated as a predicate and never as a term, to keep the traditional terminology in focus.

[48] Can one see here intimations of "relative identity", an idea that Geach developed two decades later in his (1) and (2), or does Russell have on his mind the notion of resemblance?

[49] Emphasis added here as well as in what follows.

[50] This shows Meinongian and Fregean legacies in Russell (3).

[51] Russell's theory of descriptions is geared to eliminate the opaqueness of the term and to recast the sentence to turn the term explicit.

[52] In (2) itself Russell mentions his strategy of "defining the class of men as the class of objects having the predicate human."

[53] Russell (3), Vol.1, p.57; emphasis on the original.

[54] Evidently, this way of stating the reducibility postulate (or the axiom of reducibility) - to ignore the difference between a postulate and an axiom - is at variance with the usual way in which it is stated, specially the way Russell does it in his (3). Our choice here is not motivated by a desire for novelty or originality; it was intended to bypass needless debates and polemics, and to minimize if not preclude dialectical debates. Russell did state the postulate in several ways using the conceptual battery of propositional functions, properties of different orders etc., because his aim in introducing this postulate was not just to be able to define numerical identity in terms of the logical apparatus that he erected, but a larger one of which this is only a part, namely to be able to obtain a framework required and is adequate for sustaining his no-class theory. He saw the reducibility postulate as a palatable surrogate to the distasteful assumption of the existence of classes; as he said on p.58 of his (3)

"if we assume the existence of classes, the axiom of reducibility can be proved."

Russell's interest in accepting this postulate was "largely inductive," namely that without it or something equivalent to it, far too many mathematical results - "a great mass of reasoning", as he would say, would be lost. For example, suppose that one wants to define an inductive number as possessing every hereditary property of zero; it involves unrestricted quantification not permitted by the (ramified) theory of types. The reducibility postulate helps in overcoming such problems. Of course, as Russell remarked in his Introduction to the first edition of (3) - and remarked with all seriousness and not with tongue in cheek - "the method of 'postulating' ... has many advantages; they are the same as the advantages of theft over honest toil." Incidentally, Dedekind held that this method was integral to the honest trade of mathematics. In fact this trick is as old as Eudoxous of Alexandria. That apart, in the absence of the axiom of reducibility, "strict identity" would have to be taken as a primitive idea, and - as Myhill (1) has shown - no definition of it will be available; this implies that the so-called Leibnitz's Law will be unprovable.

The variety of uses to which the reducibility postulate can be put, and its non-logical nature, together with its requirement for preserving the received body of mathematics made Russell say several not inaccurate but confusing things about the postulate. For instance, on pp.56-58 of his (3) he writes:

"The axiom of reducibility is the assumption that given any function there is a formally equivalent predicative function ..."
"In symbols, the axiom is: $(E\psi):\phi x \equiv_x \psi ! x$."
"The axiom of reducibility states that ... any property of an object belongs to the same collection of objects as those that possess some predicate."
and
"The axiom of reducibility is equivalent to the assumption that any (intentionally given) combination or disjunction of predicates is equivalent to a single predicate."

Though our purpose is not to isolate the most elegant or accurate statement of the postulate but only to note it in relation to contextually defining definite descriptions, we still may observe that it can be formulated as:

> To any predicate of an object a there exists a logically equivalent predicate which presupposes only objects of the type a.

[55] In linking thus the axiom of reducibility and his definition of identity, Russell drags in Leibnitz and his principle of the identity of indiscernibles. But this was not needed at all. Moreover, what he says about Leibnitz's principle is rather confusing. On the other hand he says that his axiom "has a certain affinity with" Leibnitz's Principle, and on the other hand he says that given the axiom, Leibnitz's principle can be proved. Myhill (1) has shown that the principle, in its generality, cannot be proved if types were to be ramified. Russell remarks also that Leibnitz's principle "supplied" the axiom - and there is a large amount of truth in this. In any case, what he seems to be taking as Leibnitz's principle - though he emphasizes that he is not interpreting him - is not strictly relevant here; so I postpone what I have to say on that to Part Two.

[56] Russell (6).

[57] p.267.

[58] In his (1) Wittgenstein argued that if our language were perfect in its syntax, not only the theory of types, but the axiom of reducibility and even identity and the theory of descriptions all would be redundant.

Chapter Three : The Poineer's Prescience

Though most of us either are or else assume that we are more intelligent than him, and hence do succeed in finding fault with his exposition, only Russell had the genius which gave him those insights which we can develop in ways of which we are convinced that they are superior to his in some marginal or even mysterious sense. Thus the need to emphasize that all those inadvertencies in Russell's writings to which I drew attention, and all the improvements I may have suggested are intended only as footnotes to his text, and do not go even to indicate an alternative to his way of absorbing definite descriptions into the syntax and the semantics of the language which he specified in the Principia. Further, in the stupendous effort that philosophers have put in all those years to understand the logical nature of definite descriptions, the fecund insights of the poineer seems to have been either ignored or simply missed. I trust that it is time to take note of some of them, at least. Of course, there are some Russell-scholars by whose standards he did not have anything original, and according to whom each of his ideas concerning definite descriptions - even his strategy of handling them - can be traced either to Frege or to Peano, in addition to Bradeley and Moore. This is what philosophical fundamentalism is; did not Whitehead say that all subsequent philosophy is a set of footnotes to Plato and Aristotle! But he at least was speaking figuratively. It is no doubt true that some of Russell's ideas, or their intimations, can be discerned in the writings of Peano and others like Bradeley, but that should not matter much, even if it were the case that Bradley decided to publish 'On Denoting' because of this. What really matters is what Russell did to them, and his insights which went into the way he exploited those ideas and strategies in the way he did. He was one of those rare philosophers who did not fall a victim to the professional delusion; each philosopher usually thinks that he is incorporating in his thought the essence of all that is important and veracious in his discipline as it has been given to him, and also that, at the same time, is absolutely original. So it is desirable to take note of his insights which enabled him to seize what was given to him and give it a new twist and a novel turn, a turn not anticipated by anyone earlier.

The first of the insights of Russell which needs to be constantly kept in mind is that not all syntactically well-formed expressions either are or else need to be semantically autonomous for a successful functioning of language be it one which is used in day to day discourse, or else one that is used in our theory-building activity. It is a commonplace that a scientist, say a physicist, relates the laws he establishes to the world only indirectly. He relates them first to some simple theoretically constructed objects and with the help of the laws relating to these analyzes the complicated behaviour of the real bodies. But because of this indirect or even partial linkage between its language and the world, physics is not worse off. And so is the case with ordinary discourse. Thus a sentence can still be indicative, for example, 'the author of Waverley is a poet' - though it contains as its grammatical subject an expression which does not indicate anything in isolation. And a sentence which is indicative in this sense of the term has a truth-value, and its truth-value is uniquely determined by sentences which it either presupposes or else entails. Thus though all sentences are not assigned truth-values directly (this is to say that in terms of the semantic relation that holds between language and the world) each sentence comes to have a truth-value. Some sentences are assigned truth-values in terms of these syntactical relationships to sentences which are unproblematic and are assigned truth-values directly on the basis of their semantic relationship to the world.

> "A phrase containing 'the' always presupposes some initial propositional function not containing 'the'."[1]

This however does not mean that it is these phrases - definite descriptions, i.e. - that stand in relation of presupposition or of entailment to other expressions. Presupposition and entailment are relations which hold between sentences, and not sentences and terms. They are relations which can be defined over a homogeneous realm of relata. This is evident from what Russell says while continuing the remark quoted above; he adds that

> "thus instead of 'x is the father of y' we ought to take as our initial function 'x begot y', then 'the father of y' means the one value of x which satisfies the propositional function"

namely 'x begot y'. Thus a sentence containing a definite description presupposes a sentence not containing it, in fact not containing a definite description at all; this is to say that a definite description functions like a reflexive pronoun, in a sense. Now, what does it mean to say, for instance, that $(\mathrm{I}x)\phi x$ presupposes say ψy? It means that a sentence containing $(\mathrm{I}x)\phi x$ as its apparent or grammatical subject cannot be true unless, among other things, it is the case that for some sentential function say ψy, ψy yields a true sentence. It may be noted here that in almost all the secondary contributions on the issue at hand it has been emphasized that presupposition leaves it open whether the presupposing and the presupposed sentences do indeed have truth-values. But the issue of sentences being truth-valueless does not arise within the framework of Russell's semantics. Thus we can say that every sentence presupposes some sentence in which definite descriptions do not occur; in the case of sentences which do not contain definite descriptions this is vacuously true. As a degenerate case we have that a sentence presupposes itself. And what we mean when we say that a sentence say ϕ entails another sentence say ψ, is that if ϕ is true then ψ must be true, and if ψ is false then ϕ must be false. Thus each sentence entails some sentence; again as a degenerate case we have that a sentence entails itself. This means that the class of sentences is closed under presupposition as well as entailment[2]. And just as the truth of a presupposition of a sentence ϕ has the necessary intimation of its truth, the falsity of what it entails has the sufficient intimation of its falsity. Thus 'is true' is neither indeterminate, nor is it a partially defined predicate; it can be well-defined and there are no truth-value gaps. The truth-value of $\chi(\mathtt{I}x)\phi x$, for instance, can be worked out from the truth-values of what it presupposes and what it entails. This means that the syntactical and the semantic structures of language are inductive in their nature, and hence the truth-value of a sentence of a language is linked to the truth-values of various other sentences of that language.

As there is a vast and confusing literature on what a sentence containing a non-indicative or an incomplete expression as its (apparent) subject means, it is desirable to pay some attention to the notion of a non-indicative term and the notion of an incomplete expression with the intention of first differentiating the senses in which Russell used them from the senses in which others understood them, and secondly to note the legitimacy of the way in which Russell used them. A sentence in which an expression with 'the' as its prenex figures as the apparent singular subject term can take either of the following canonical forms.

(I) $x = (\mathtt{I}z)\phi z$

or

(II) $\chi(\mathtt{I}x)\phi x$.

This is to say that it can either be an identity sentence or a predicative sentence with a definite description as its subject term. Consider first a sentence of the form (I), say 'x is the father of y', to take Russell's own example. (There is no need to object to our taking into account sentential functions rather than sentences while illustrating the point. Once we are clear about the syntactical and semantic relationship between sentences and sentential functions, no harm will come of our using sentential functions in the way we will be using here. In fact it will shorten our explanations, for we need not resort to needless beaten track details each time. Thus we may talk of the truth of ϕx to mean the truth of a sentence which it yields for a suitable substitution of the variable, thus say ϕa. This is only indicative of our preference for laziness over boredom.) This sentence, says Russell, presupposes another sentence, say 'x begot y' where x as it occurs in the former sentence is the only value of x as it occurs in the latter sentence. The former sentence is an identity, and is of the form $x = (Iz)\phi z$. Thus y as it occurs in 'x is the father of y' disappears from the realm of the primitive terms; it does not figure either as or in the subject part of the sentence, and finds a mention in the predicate part[3].

What Russell says is that $x = (Iz)\phi z$ cannot be true unless some sentence (not necessarily different), say Bxy, is true and the only value of x as it occurs in Bxy is the one value of x in 'x is the father of y'. Now, $x = (Iz)\phi z$ is not an incomplete sentence; it is a well-formed expression which is a full-blooded sentence. Only $(Iz)\phi z$ is incomplete, and only

> "phrases (other than propositions) containing <u>the</u> (in singular) are"

incomplete, as Russell would put it. (In the foregoing as well as in what follows there is a lot which needs to be taken without "begrudging a pinch of salt".) All this explanation holds good as heuristic; much of it needs to be ignored when a modicum of understanding of definite descriptions is gained, this to say that when proficiency in the language of the <u>Principia</u> is acquired. In this language, $x = (Iz)\phi z$ is neither an identity sentence, nor a complete sentence. It would be an identity sentence only if $(Iz)\phi z$ is a proper description, and $x = (Iz)\phi z$ would be a complete sentence only if the scope indicator is prefixed to it; it is thus a short for $[(Iz)\phi z\{x = (Iz)\phi z\}]$. Further, $x = (Iz)\phi z$ is a sentence

> "which may be true or may be false, but is never trivial."[4]

This means that $x = (\iota z)\phi z$ is not analytic; it is contingent and is an identity sentence whose truth-value is the same as that of $(Ex)\{x = (\iota z)\phi z\}$. This sentence will be true if the variable x in it were to take $(\iota z)\phi z$ as its value. Thus the truth of an identity sentence in which one of the terms is a phrase with prenex 'the' in singular, a definite description that is, presupposes a sentence in which the variable of which the definite description in question is a value is existentially bound. Russell could have brought out this by taking the sentence $(Ex)\{x = (\iota z)\phi z\}$ itself as an example; if he took Bxy as an example, it could be that he himself was, or else thought that other may be, under what Russell called

> "the silly delusion that if an x and an a (occur in) the same proposition, the x cannot take the value a."[5]

In any case what Russell wanted to impress upon his readers was that, as the truth-value of Bxy and (Ex)Bxy are the same; if $x = (\iota z)\phi z$ is true, Bxy must be true, and hence (Ex)Bxy must be true. Suppose $x= (\iota z)\phi z$ is true, then because $B(\iota z)\phi z$, y is true, Bxy must be true.

The case is the same with $\chi(\iota z)\phi z$, this is to say that with a predicative sentence in which a definite description occurs as a subject term - 'the author of Waverley is a poet', for instance. Evidently, that the x who wrote Waverley is a poet cannot be true unless there is somebody who is in fact a poet, and that he be the one who wrote Waverley. This means that if it is true that the author of Waverley is a poet, then it must be true that there is some y such as that y is poet - thus $(Ex)\chi y$, to put it in the language of the Principia - and it is also true that $y = (\iota x)Wx$. It may be noted that this is not the same as saying that $(\iota x)Wx$ is a value of the variable y as it occurs in $(Ex)\chi y$, which may not be true. Obviously, from that somebody is a poet, it does not follow that the one who wrote Waverley is a poet. This shows first 'that $(\iota x)\phi x$ is not a value of y', and secondly that identity (pronomial identity, i.e.) is built into definite descriptions. If $(\iota x)Wx$ were a value of the variable y in $(Ey)\chi y$, then $y = (\iota x)Wx$ would be a nomial identity sentence, which it is not. y as it occurs in $y = (\iota x)Wx$ is a pronoun. A significant use of a pronoun presupposes a similar use of a noun, and an indication to the effect that the pronoun in question is proxying the noun. Evidently the story cannot begin with 'He had three sons', but with 'Once upon a time there was a Sultan of Isfahan' followed by the sentence about his moderate siring. Here one might demand a justification of our earlier remark, namely that the point Russell wanted to highlight could as well have been brought forward by taking $y = (\iota x)Wx$ itself as an example. Because y is a pronomial term in $y = (\iota x)Wx$, it will not be possible to move from it to $(Ex)\{y = (\iota x)Wx\}$ unless it is shown that it indeed is a proxy for an occurrence of y which is nomial, this is to say that there is a sentence in which y occurs nomially. Thus, for instance, unless it is the case that somebody is something it cannot be said that he (in fact) is the author of a set of boring novels. And to say that somebody is something is the same as to say that he has such and such property, or that he instantiates that property. Only after securing that we can say, truly or falsely, that he is the one who instantiates some other properties too.

> "Thus in all such cases, the existence of the (grammatical) subject $(\iota x)\phi x$ can be analytically inferred from any true proposition having this grammatical subject."[6]

Any sentence in which y occurs nomially is sufficient for the purpose at hand[7], hence we can as well let it be $(Ex)Wx$, that is 'somebody wrote Waverley' - or, as by now we are infected by the Principia, 'something wrote Waverley'. Given that something wrote Waverley, we can say that it alone wrote it, and that it is one which is such and such also. Thus $(Ey)\{y = (\iota x)Wx\}$ presupposes $(Ey)Wy$. The same holds in the case of a predicative sentence containing a definite description as its grammatical subject, for example 'the author of

Waverley is a poet' or schematically $\chi(\mathtt{I}x)\phi x$. If this is true, so must $(\mathtt{E}y)\chi y$ and $y = (\mathtt{I}x)\phi x$ be; here ψ may even be χ.

> "Thus although '$(\mathtt{I}x)\phi x$' has no meaning by itself, it may be substituted for y in any propositional function Fy, and we get a significant proposition though not a value of Fy."[8]

This needs to be taken into cognizance in order to be able to correctly understand what Russell means when he says that a sentence of the form $\chi(\mathtt{I}x)\phi x$ - be it an identity sentence or a predicative one - presupposes $(\mathtt{E}y)\chi y$[9].

Now suppose that $\chi(\mathtt{I}x)\phi x$ is true. Then $(\mathtt{E}x)\chi x$ must be true, and it follows from this that $y = (\mathtt{I}x)\phi x$ is true. Further, that $y = (\mathtt{I}x)\phi x$ can easily "be shown to be equivalent", as Russell would say, to $(x)(\phi x \equiv x = y)$. As this holds for any y, it follows that $(y)(x)(\phi x \equiv x = y)$. Thus if $\chi(\mathtt{I}x)\phi x$ is true, then it follows that $(y)(x)(\phi x \equiv x = y)$ is true, and thus presupposes $(\mathtt{E}x)\phi x$. So if it is true, so must $(\mathtt{E}x)\phi x$ be. Thus Russell sums up:

> "it is plain that any statement apparently about $(\mathtt{I}x)\phi x$, requires (i) $(\mathtt{E}y)\phi y$ and (ii) $(y)(x)(\phi x \equiv x = y)$."[10]

Then when $\chi(\mathtt{I}x)\phi x$ is true, $(\mathtt{E}y)\{\phi y . (y)(x)(\phi x \equiv x = y)\}$ is true, which is the same as to say that $(\mathtt{E}y)\{\phi y . (x)(\phi x \equiv x = y)\}$ is true. Russell thus comes to the conclusion that $(\mathtt{E}y)\phi y$ and $(y)(x)(\phi x \equiv x = y)$

> "together are equivalent to $(\mathtt{E}x)[\phi y . (x)(\phi x \equiv x = y)]$ which is defined as $E!(\mathtt{I}x)\phi x$, and must be a part of what is affirmed by any proposition about $(\mathtt{I}x)\phi x$."[11]

To take into account the presumed entailment, suppose that ϕ is required for ψ, then ϕ should naturally follow from ψ. Thus $\chi(\mathtt{I}x)\phi x$ entails $(\mathtt{E}y)\{\phi y . (x)(\phi x \equiv x = y)\}$. The upshot of the foregoing then is that the indication of $(\mathtt{I}x)\phi x$

can be discerned in $\chi(\mathrm{I}x)\phi x$, which in turn can be mapped onto the indication of

$$(Ey)\{\phi y . (x)(\phi x \equiv x = y)\}.$$

In so far as the indication of this sentence is precisely determined by the variables that figure in it, all this amounts to reducing indication as such exclusively to individual variables, bound variables; Quine smelled this as naturally as a cat would smell a rat. The first important insight of Russell is encoded in this.

What emerges from the above discussion is that a sentence with $(\mathrm{I}x)\phi x$ as its (grammatical or apparent) subject term presupposes existence, and entails uniqueness - existence and uniqueness as captured by individual variables and identity in the canonical language. And, as uniqueness itself presupposes existence, it follows that it entails existence as well as uniqueness. Now, what a sentence entails, in the restricted sense of the term in which it is being understood here, concerns only with the indication of the sentence, and not whether it instantiates the predicate figuring in the sentence[12]. This is to say that the content of the sentence $\chi(\mathrm{I}x)\phi x$ is not exhausted by what it entails, which as has been noted constitutes only one half of it. The other half of its content is in that the (bound) variables figuring in the sentence entailed by $\chi(\mathrm{I}x)\phi x$ do instantiate the predicate χ [13]. The mapping of $(\mathrm{I}x)\phi x$ onto $(Ey)\{\phi x . (y)(\phi y \equiv x = y)\}$ is thus only a preparation for reducing $\chi(\mathrm{I}x)\phi x$ onto $(Ex)\{(y)(\phi y \equiv x = y) . \chi x\}$.

This reduction shows how <u>the content of a sentence is exhausted only by a conjunction of what it presupposes and what it entails</u>; thus, what a sentence presupposes as well as what it entails form parts of the content of that sentence. When a sentence is asserted, the logical relationship between these two parts is asserted[14].

The reduction in question is syntactical in its nature, but in effecting it some semantic information is taken into cognizance. This however should not deprive it its priority; and without such a reckoning of semantic information, this syntactical reduction would be purely ad hoc or arbitrary[15]. Because this reduction is syntactical, assignment of truth-values to sentences like $\chi(\mathrm{I}x)\phi x$ too is a syntactical affair. In so far as truth-value is determined in terms of the truth-value of the components, truth becomes a syntactical notion. Assignment of a truth-value to a sentence like the one in question on the basis of its syntactical form, and the truth-value of the sentence to which it is related, are matters which do not require us to know how atomic syntactical units are assigned truth-values. All that is being said here is that if $\chi(\mathrm{I}x)\phi x$ is true, then

$$(\mathrm{E}x)\{(y)\,(\phi y \equiv x = y)\,.\,\chi x\}$$

must be true, and if this is false, then $\chi(\mathrm{I}x)\phi x$ must be false. Of course we know how to ascertain the truth-value of that sentence; we know also that it is either true or false. Thus $\chi(\mathrm{I}x)\phi x$ must be either true or false. Now if it is false, then $\sim\chi(\mathrm{I}x)\phi x$ must be true, because 'truth' here is a normal notion, and the connectives are standard.

However, this leads to an irritant. As we noted earlier in Chapter two, $\sim\chi(\mathrm{I}x)\phi x$ reduces either to

(I) $\sim(\mathrm{E}x)\{(y)\,(\phi y \equiv x = y)\,.\,\chi x\}$,

or to

(II) $(\mathrm{E}x)\{(y)\,(\phi y \equiv x = y)\,.\sim\chi x\}$.

This means that to say $\sim\chi(\mathrm{I}x)\phi x$ is true if $\chi(\mathrm{I}x)\phi x$ is false necessitates accepting that the denial of $\chi(\mathrm{I}x)\phi x$ reduces to a disjunction. This indeed is what Russell's analysis amounts to, and this is natural too, it seems[16].

If $\sim\chi(\mathrm{Ix})\phi x$ is true, it may be the case that $\sim(\mathrm{Ex})\phi x$. This is to say that $\sim\chi(\mathrm{Ix})\phi x$ may be taken as $\sim\{\chi(\mathrm{Ix})\phi x\}$. If $\sim(\mathrm{Ex})\phi x$ is true, which is the same as to say that if ϕ is an uninstantiated predicate, it cannot be one which is uniquely instantiated either. Then $\sim\chi(\mathrm{Ix})\phi x$ entails $\sim(\mathrm{Ex})\{(y)(\phi y \equiv x = y)\}$, and reduces to (I) above. It may be the case that $(\mathrm{Ex})\phi x$ is true, but if it false that $(y)(x)(\phi x \equiv x = y)$, then $(\mathrm{Ex})\{(y)\ (\phi y \equiv x = y)\}$ would be false, and hence $\sim(\mathrm{Ex})\{(y)(\phi y \equiv x = y)\}$ would be false[17]. Then in this case too $\sim\chi(\mathrm{Ix})\phi x$ entails $\sim(\mathrm{Ex})\{(y)(\phi y \equiv x = y)\}$ and reduces to (I). Finally, $\chi(\mathrm{Ix})\phi x$ may be false, and hence $\sim\chi(\mathrm{Ix})\phi x$ is true; and this may be the case not because $(\mathrm{Ex})\{(y)(\phi y \equiv x = y)\}$ is true, but because χx is false. If χx is false, then $\sim\chi x$ is true, which is the same as to say that $(\mathrm{Ex})\sim\chi x$ is true. In so far as this is the presuppositional part of the content of $\sim\chi(\mathrm{Ix})\phi x$, it amounts to treating $\sim\chi(\mathrm{Ix})\phi x$ as $(\sim\chi)(\mathrm{Ix})\phi x$. The entailment part of its content thus would be $(\mathrm{Ex})\{(y)(\phi y \equiv x = y)\}$. Thus in this case $\sim\chi(\mathrm{Ix})\phi x$ reduces to $(\mathrm{Ex})\{(y)(\phi y \equiv x = y).\sim\chi x\}$. This means that to say $\sim\chi(\mathrm{Ix})\phi x$ is true is the same as to say that the disjunction with (I) and (II) as its two disjuncts is true; thus if it is true,

$$\sim (\mathrm{Ex})\{(y)(\phi y \equiv x = y)\} \vee \sim\chi x$$

is true.

Some Indispensable Exegesis:

This digression is necessitated by the secondary literature which is terribly confusing on some teasing if not intriguing issues with which Russell was concerned, specially issues relating to the notions of presupposition, truth-value, a sentence's being a part of the content of another sentence, and above all the notion of definition itself. To keep my discussion manageable, I shall restrict it to the most essential points in the relevant writings of Strawson,

Linsky, and Consuegra[18], and relegate discussion of secondary issues to footnotes and Appendix. 14.01 of the Principia reads as follows:

$$\{(\mathtt{I}x)(\phi x)\}\,.\,\psi(\mathtt{I}x)(\phi x).=.(Eb):\phi x.\equiv_x.x = sb:\psi b \text{ Df}.$$

which in a more familiar notation would read as:

$$\psi(\mathtt{I}x)\phi x =_{df} (Ey)\{(x)(\phi x \equiv x = y)\,.\,\psi y\}.$$

Moving towards 14.02 Russell pays attention to the logical relationship between the indication of the presupposition of a sentence in which the subject term has a prenex 'the' in singular, and the indication of what a sentence entails. Suppose that the definitum of 14.01, that is the left sentence in it, is true; then it is true that $y = (\mathtt{I}x)\phi x$, which is the same as to say that the indicatum of the definitum and the indicatum of the definiens are the same. Now, suppose that $y = (\mathtt{I}x)\phi x$ is false, then the definitum is false; but it will be so either because $y \neq (\mathtt{I}x)\phi x$ is true, or else because $\sim\{y = (\mathtt{I}x)\phi x\}$ is true. These two sentences are not the same because

$$y \neq (\mathtt{I}x)\phi x.=.(x)(Ez)(\phi z \equiv x = y)\,.\,y \neq z,$$

and

$$\sim\{y = (\mathtt{I}x)\phi x\}.=.(x)\sim\{(Ez)(\phi z \equiv x = y)\,.\,y = z\}.$$

Of these two, "the first necessarily implies $(x)(Ez)(\phi z \equiv x = z)$," while "the second does not" - to put it in Russell's own words[19]. Thus Russell remarks

> "(W)e put
> 14.02 $E!(\mathtt{I}x)\phi x.=.(Eb):\phi x.\equiv.x = b \text{ Df}$."

Consider now what he writes on p.68, Chapter III of the Introduction to the Principia, namely,

> "... it is plain that any statement apparently about $(\mathrm{I}x)\phi x$ requires (1) $(\mathrm{E}x)\phi x$ and (2) $(\phi x . \phi y). \rightarrow ._{x,y} x=y$; here (1) states that at least one object satisfies ϕx, while (2) states that at most one object satisfies ϕx. These two together are equivalent to $(\mathrm{E}c):\phi x . \equiv ._{x} x = c$ which is defined as $\mathrm{E}!(\mathrm{I}x)\phi x$."

Thus

> "$\mathrm{E}!(\mathrm{I}x)\phi x$ must be a part of what is affirmed by any proposition about $(\mathrm{I}x)\phi x$."[20]

What are the definienda respectively of the definitions 14.01 and 14.02? There seems to be no problem concerning the definiendum of 14.01; without doubt it is $\psi(\mathrm{I}x)\phi x$. It is a definition which conforms to the requirements of being a mathematical definition which Russell was articulating[21]. The problem thus boils down to (1), specifically whether it entails that $(x)(\mathrm{E}y)(\phi x \equiv x = y)$ is a part of the content of $\psi(\mathrm{I}x)\phi x$ or not[22]. Earlier I cited two passages in which Russell is explicit that $\psi(\mathrm{I}x)\phi x$ has a presupposition, and I tried to unpack some relevant passages to point out what that presupposition was. Thus if we find Strawson insisting on that Russell was oblivious of the presupposition of a sentence containing a definite description as its subject term, it could perhaps be due to the reason that he thinks that failures of presupposition necessarily lead to truth-value gaps, which Russell did not admit. Strawson's contention will come to hold if "truth" and "falsity" are not bipolar, and are not interdefinable. As Russell says, in his reply to Strawson, he finds

> "it more convenient to define the word 'false' so that every significant sentence is either true or false."[23]

In all these years Strawson could not show that 'truth' and 'falsity' are not bipolar; nor could he justify that a semantic theory is inutile, and is unfair to our representation of the world, unless it contains two independent definitions, one of truth and another of falsity. Thus his accusation, that Russell maintained that a part of the content of the definiendum is indeed the definiens because he was unaware of the relation of presupposition (as such) holding between them, does not stand the test of reason as fairness to others. However, Russell did hold what Strawson says that he did, but for a different reason. There are too many Russell-scholars who hold that Russell did subscribe to the view that the definiendum and the definiens are related in this way. Consuegra, for instance, was firm in asserting that "Russell never held it," and he invokes the authority of Linsky in his support[24]. In reply to him, and to Linsky if he has been correctly represented by Consuegra, I may just quote the following passage from Russell (3):

" ... E!(Ix)ϕx is, by definitions (i.e. 14.01 and 14.02) part of ψ(Ix)ϕx."[25]

This one sentence should be sufficient as a reply to Strawson and Consuegra; but if I am induced to harp on it a little more, it is not due to lack of prudence, but because it is needed to indicate my stand on an important point in Russell-studies, all the more so as Consuegra writes:

> "In our day many authors try to reach the "truth" about a certain historical questions by means of personal judgment and starting from very limited relevant material, but I try to show in this article (and in general in my 1987a) that usually it is necessary to handle a greater background, including unpublished material and study of influences."[26]

I am one of those numerous authors, and fortunately I am in the not so bad a company as that of Strawson and other half-literates. So I start from, and shall not go beyond, "very limited relevant material," as Consuegra would say. What is of importance is that it is relevant and not otherwise as most of the unpublished material seems to be[27]. What is required is a careful reading of the relevant material, and not a statistical analysis of the unpublished writings which the writer did not intend to publish. This assumes special significance in the case of writers like Russell, who never lost a moment in pushing their scripts to publishers, nor are deprived of access to them. With reference to such authors sticking to their published works seems to be the sanest policy. If Russell thought that he had something important to say in some of those writings he would have certainly published them. Not observing this much of prudence is tantamount to prying on others' waste-paper baskets, a sickly habit encouraged and cultivated by affluent academies and academics. Moreover, handling unpublished material need not issue in greater philosophical illumination, or even historical accuracy. I mention the latter as Consuegra's noble aim is to reconstruct "the general evolution of the historical Russell." I do not intend to do that, for the historical Russell is not the living Russell; thus as I see what is required is to isolate the surviving Russell from the dated Russell. Consuegra is indifferent to this task, and on his intended goal he fares badly. In any case, as polemics rarely result in philosophically interesting results, we may return to the point at issue, and to the point where we left it.

What is the definiendum, and what is the definiens of 14.02? The answer, obviously, is: either of the sentences occurring in it - it all depends upon what one wants to do with it, and the context in which he needs to invoke it. I am not imputing either vagueness or unawareness to Russell, nor am I reading into or interpreting and paraphrasing what he did say. He is admirably clear on this point; 14.204 of the Principia reads as:

E!(Ix)ϕx.≡.(Eb).(Ix)ϕx ≡ b [28].

E!(Ix)ϕx is neither a sentence of the language to which the definiendum of 14.01 belongs, nor is it a sentence of the language to which the definiens of 14.01 belongs. E! is not a predicate of either of these languages. Thus 14.02 cannot be given the same place as 14.01 which was geared to incorporate a theoretical advance in giving a precise formulation of the content of a philosophically confusing sentence; it was intended to go as a logico-mathematical definition in the sense of the term specified in Chapter Two. In contrast 14.02 does not seem to do anything of that sort; it suggests itself to be a definition in a degenerate or a charitable sense of the term, this is to say that as a record of an abbreviation. Russell uses throughout Chapter 14 of the Principia the apparent definiendum of 14.02 as an abbreviation of its apparent definiens. Of course, he rallies it for a purpose other than shortening sentences of the language to which the definiens of 14.01 belongs. Going by the "global vision of Russell's work" - to borrow a facile phrase from Consuegra - we may note that Russell uses E!(Ix)(ϕx) as an auxiliary sentence, and E! as an auxiliary predicate, with some transitory role and explanatory significance. Its role evidently is in specifying the way in which the truth-value of ψ(Ix)ϕx can be fixed in terms of the truth-value of (Ex){(y)(ϕy ≡ x = y) . ψx}. Earlier we noted how the truth-value of ψ(Ix)ϕx is conjunctive, and its falsity is disjunctive, this is to say that how unless ψ is an instantiated predicate, and (Ix)ϕx be an instantiation of it, ψ(Ix)ϕx cannot be true, whereas it can be false either when (Ix)ϕx does not instantiate ψ, or else when (Ix)ϕx does not exist. Now, only

> " ... when (Ix)ϕx exists, it has any property which belongs to everything. This does not hold when (Ix)ϕx does not exist, for example the present king of France does not have the property of being either bald or not bald."[29]

Suppose that (Ix)ϕx does not exist. Then, if Strawsonian or van Fraassenian relation of presupposition holds between the definiendum and the definiens, E!(Ix)ϕx will be truth-valueless; but this sounds odd, as that a non-existent is existent is an explicit contradiction and hence false, as much as that an existent is existent is a tautology and hence true[30]. Such expressions can always be eliminated in terms of the mechanism of quantification. That precisely is what Russell does.

What does it mean to say that something exists? Whatever may be the way in which one unpacks the notion of existence, it must include that it has this much built into it, namely that an existent has some property and thus there are no uninstantiated properties[31]. (There may be predicates without correlated properties, but that is a different matter.) Our language succeeds in giving a description of the world on the assumption that there are neither bare individuals - "coat-hangers," as Russell would call them - nor uninstantiated properties. So to exist, besides being any other thing one's fancy disposes him to entertain it to be, is to be an instance of property. This minimal sense of existence is logically maximal; and it can be exhaustively and exactly characterized within the first order quantification theory, thus the language of the Principia. To say that a property is instantiated is to say that the corresponding expression is a predicate such that the sentence $(Ey)\phi y$ where ϕ is that predicate, is true. Similarly, to say that $(Ix)\phi x$ exists is the same as to say that it is the value of a variable say y in some sentence of the form $(Ey)\psi y$, where ψ may be ϕ itself; this is the same as to say that $y = (Ix)\phi x$ is true. This enables us to assign truth-values to sentences without leaving any gaps, and thereby turn the very notion of a sentence a precisely determined one.

Let us now pay some attention to the second fecund insight of Russell. What emerged and became transparent from the foregoing is that existence is taken care of by bound variables, and that uniqueness can be captured without residues with bound variables, and identity. Because of this, existence is redundant, $E!$ is an inutile predicate in the language of the Principia. Delinking it from its polemical context, we may note that Geach has admirably netted this insight of Russell in the following passage.

"It makes absolutely no logical difference whether we say "There is something that is F" or "There exists something that is F"; "exists" is merely a shade more formal than "is". It ought not to be necessary to say this; but it is necessary in view of things some Oxford Philosophers say about "exists". Some of these ... besides happening to be false, could not possibly be philosophically relevant ... Russell has repeatedly pointed this out; but these Oxford Philosophers despise Russell and do not read him."[32]

Of course, the idea is all too pervasive in the writings of Frege, and forms part of his semantics, though he explicitly stated it in a passage with which Russell could not have been familiar because even in its original version it was published only recently. Stating it semantically, Frege remarked:

" ... if 'Sachse exists' means 'the word 'Sachse' is not an empty sound, but designates something then it is correct that the condition be fulfilled. Moreover, this condition is not a new premise, but the obvious precondition of all words. The rules of logical analysis assume that the words used are not empty, that sentences are expressions of judgments, that one is not merely playing with words."[33]

Russell is attempting to capture this brute fact of semantics in syntactical terms. In so far as a quantifier is a variable binder, the intended goal requires identity[34]. How vexed Russell was with this formidable problem can be surmised from Ramsey's communication to Wittgenstein who thought that Russell's analysis of identity was his greatest folly[35]. (That these two are among Russell's intellectual intimates hardly needs a mention.) On 12 November, 1923 he wrote to Wittgenstein:

"Have you noticed the difficulty in expressing without = what Russell expresses by (Ex)(Fx . x = a)?"[36]

Writing again on the 20th of the following month, he added:

" ... I am afraid you will think my difficulty silly. I didn't think there was a real difficulty about (Ex)(Fx . x = a) i.e. that it was an objection to your theory of identity, but I didn't see how to express it, because I

> was under the silly delusion that if an x and an a occurred in the same proposition the x could not take the value a. I had also a reason for wanting it not to be possible to express it. But I will try to explain it all in about a fortnight from now, because it might help me to get clearer about things."

We do not have any information suggesting that Ramsey did write the promised letter, but this much seems to be a fair bet, namely that Ramsey appreciated the difficulties involved in absorbing existence into quantification theory, and also that he noticed how Russell overcame those difficulties in terms of his theory of identity. When quantifiers are given existential reading, there will be no way in the unextended quantification theory to formulate a true negative singular existence sentence, for instance 'a does not exist' where a is an "empty" singular term. The difficulty could be easily surmounted by treating 'being-called-by-a-name' as a property of the object thus called. But this way out must have sounded odd to Russell with his robust sense of reality and a firm commitment to realism[37]. It is implicit in the very notion of a property that it determines the nature of any object that has it, and the nature of an object is not affected if it is called by name rather than the other. This could be one of the reasons for Russell's not recommending the elimination of proper names altogether as Quine subsequently came to advocate; the other reason could be his belief, which he shared with Frege, that there should be at least one proper name in a language if it were to be of any use for representing the world. More importantly, by invoking identity, distinct true negative existential singular sentences - for example 'Pegasus does not exist' and 'Round square does not exist' - can be formulated without producing anomalies, and without collapsing the diversity of the singular terms in them. Obviously,

~(Ex)(x = Pegasus),

and

~(Ex)(x = Roundsquare)

will not be logically identical; they will not be identical because two things that are diverse from a given thing need not be identical. Perhaps, this is why Russell accepted diversity as a primitive relation[38]. And this is in consonance with Russell's view of the nature of a logical theory and its construction which he succinctly stated in his (2).

> "A logical theory may be tested its capacity for dealing with puzzles, and it is wholesome plan, in thinking about logic, to stock the mind with as many puzzles as possible, since these serve much the same purpose as it is served by experiments in physical science."[39]

Russell took care that the test to which his logical theory was put itself satisfies the criteria of legitimacy. Consequently, his theory meets the two demands, namely the Occhamite demand and its counterpart. His theory conforms to the Law of Parsimony, that is that syntactical categories must not be multiplied beyond necessity, as well as the Law of Miserliness, namely that syntactical categories should not be reduced below sufficiency, where sufficiency is understood in terms of the requirements of a representation of the world (which, as a robust sense of reality would have it, was antecedently given.) Thus Russell would say:

> "Logic, I should maintain, must no more admit of a unicorn than zoology can; for logic is concerned with the world, just as truly as zoology, though with more abstract and general features."[40]

The crucial term here is 'general'; what Russell is pointing at is too familiar to us now, namely that quantification theory has in it the necessary and sufficient rubrics for capturing the world in its generality[41]. And to capture it in its specificity, identity and singular terms need to be brought in, Thus quantification theory extended suitably provides all the rubrics needed for representing the world. These requirements, in any case, are three in number, first that every subject term must be indicative, secondly that each term be constant or rigid in its indication[42], and finally that the terms indicate uniquely; this is to say that if a term is singular this fact should be evident from the sentence in which it figures as a subject term[43].

Thus, for Russell, a logical theory comprises of three nested parts, first a well-defined set of sentences, secondly a system of relations holding between sentences by means of which it can be decided for a given pair of sentences whether one of them follows from the other, and finally a certain criterion of truth yielding the mechanism first to assign truth-values to a subset of sentences of the language, and secondly to assign truth-values to the remaining sentences. These parts are nested in the sense that the third of these is dependent on the first two, and the second is dependent on the first. For instance, truth-value determination of a sentence depends upon whether it is an atomic sentence or a compound or complex one, for if it is of the latter sort, then its truth-value will be a function of the truth-values of the sentences of which it is a compound; it could even be a function of the truth-value is derivative of the truth-values of the sentences from which it follows, or of the truth-values of the sentences which follow from it. And whether a sentence follows from another is contingent upon their respective logical (this is to say that, syntactical) forms. In exhibiting the logical form of a sentence, and attempt is to be made to mirror its grammatical form an understanding of which is left to the native proficiency of the analyst. In parsing a sentence one comes to exhibit its logical form, and more importantly, or comes to reduce the grammatical form of a sentence to its logical compoundness.

Thus, a logical theory, for Russell, does not, nor is it intended to, replace ordinary language. On the other hand, it is geared to explain and exhibit how ordinary language succeeds in representing the world, or how we do succeed in representing the world through it. If he paid enough attention to sentences of ordinary language - in fact, he started contracting each part of his logical theory with a meticulous parsing of sentences of ordinary language.

But, like Frege, he too wanted to purge it form those features of it which bring in glaring oddities into our representation of the world through it. As Frege did say in a passage with which Russell must be familiar,

> "(I)t is customary in logic texts to warn against the ambiguity of expressions as a source of fallacies. I deem it at least as appropriate to issue a warning against proper names that have no nominata. The history of mathematics has many a tale to tell of errors which originated from this source. The demagogic misuse is as close (perhaps closer) at hand as in the case of ambiguous expressions. 'The will of the people' may serve as an example in this regard; for it is easily established that there is no generally accepted nominatum of that expression. Thus it is obviously not without importance to obstruct once for all the sources of these errors, at least as regards their occurrence in science."[44]

That the human kind is genetically endowed with the capacity to generate combinations like 'the present king of France', and 'The will of the People', and produce them according to rules that are permitted and universally accepted, is not denied either by Frege, or by Russell. What indeed needs to be denied, and what they deny, is that by producing an expression an object which is a part of the world in some well-definable and legitimate sense of the term, and is its nominatum, will have been produced. This is to say that they are holding that the rules of sign-production are not rules of individuation.

The motive behind this denial is not to prune the world; this should be evident from the expression 'the robust sense of reality' itself. Whatever is - as Plato would say - is an empirical matter; what is, cannot be wished away, nor can we whistle objects into existence. What exists needs to be noticed in observation and captured in empirical knowledge. The philosophically enticing issue is about what needs to be hypostasized in order to account for, or represent, what is. Russell's ultimate answer to this issue is: nothing. Of course, we do occasionally find the need - maybe for reasons of convenience - to drag in expressions of the form $(\hat{x})\phi\hat{x}$, or $(\mathrm{I}x)\phi x$ while accounting for the world as it is. We may even use those expressions as apparent subject terms of sentences in the indicative mood; this does not mean that each sentence of that sort is hooked to the world in the same way as are sentences with full-blooded proper names as their subject terms are. That they are hooked to the world cannot be doubted; but they are so only through their (logical) relationship to sentences whose subject terms have well-defined correlates in the world. This obviously amounts to reducing informativeness to indication, and defining indication in terms of variables and quantifiers. As Russell would say:

> "Thus every proposition that we can apprehend, i.e. not only those whose truth or falsehood we can judge of, but in all that we can think about, all the constituents are really entities."[45]

These entities are indicated by variables which specify the topic of our concern towards which the rest of the sentence in which they occur is directed. In showing how this obtains, Russell attempted to solve two basic issues in philosophy, namely first how can we gain knowledge of things with which our cognitive apparatus is incapable of interacting? After Russell, we now have a plausible explanation of the way in which knowledge about things which elude our cognitive faculty can be logically related to our knowledge gained in terms of our cognitive faculty; and we now are in a position to legitimize our belief that a thing knowledge of which cannot be so related does not form part of the world. Thus logical theory is not only an instrument which enables us to represent the world, but is an instrument of discovery as well!

In view of the fact that indication is exhaustively characterized only in terms of variables and quantifiers, it emerges that the indication of a singular term can be explained (away) in terms of variables, quantifiers, and identity. What this means is that the theory of singular terms, as Quine reaffirmed later during the last five decades, and van Fraassen hammered at during the sixties, is independent of quantification theory, and that a variable is not a place-marker for singular terms; thus a generalized sentence, either the universal $(x)\phi x$ or the particular $(Ex)\phi x$, is neither a logical product nor a logical sum of singular sentences, and more importantly quantifiers are ineliminable tools of representation[46], even if there were available to us an infinite number of singular terms.

As Grattan-Guinness informs us, in 1906 Russell started thinking in terms of the substitutional interpretation of quantifiers, but soon came to abandon it as he realized that it was incompatible with the correspondence theory of truth to which he, as a realist, was firmly committed[47]. Russell could not stomach the existence of non-facts, or "objective falsehoods" which false sentence require, and which true negative existentials[48], like 'Pegasus does not exist', need. As if he were heeding to the advice of Plato in the Sophist when he remarked:

> "keep your mind from that line of enquiry, for never will you show that non-being is,"

and realizing how Meinong, and following him he himself for sometime, got into problems by ignoring this advice, Russell noted the conceptual need to shift the very notion of negativity - negative propositions, negative facts, non-existent entities, non-being ... - from the realm of ontology to the sphere of language, and its mode of representation. This is the import of his remark, in his reply to Strawson, that he prefers to define falsity in terms of truth[49] which he understands as correspondence with he world. The assumption of this design, at once simple and grand, may not be either transparent or uncontested; nor are their consequences universally palatable. But so is the case with any known alternative to Russell's theory. In any area in which chaos reigns, little is gained upon, and new methods are eagerly seized upon to conjure or conjuncture new solutions to the problems confronted there, we can legitimately believe that we have hardly begun, and that all those diverse solutions are to be considered as alternative paths along which further moves need to be made to find out which path leads to the intended goal, and which path leads to a dead-end. This is precisely why we need to take Russell's theory as one of the on-going programs of research and not a historical curiosity. Further, the diversity of solutions to the issue at hand may be instructive of the fact that issues are not yet quite ripe to be stripped off from metaphysics or ontology, and form their own science even if the time is long past when such issues were exclusively the domain of the metaphysician's free speculation.

Notes

[1] Russell (3), Introduction to the first edition; emphasis added.

[2] This breaks down under the doctrine of "elementary propositions" in Wittgenstein (1); this doctrine does to deductive logic what Humean skepticism did to inductive reasoning.

[3] Perhaps, Wittgenstein was pointing to this when he wrote to Russell in 1914 that the variables in definite descriptions do not figure in their normal senses. See Wittgenstein (3).

[4] Russell (3), Chapter III, p.67.

[5] Wittgenstein (4), letter dated 20th December, 1923.

[6] Russell (3), Chapter 14, p.175; emphasis added.

[7] This shows how Russell's enquiry is completely extensional, and moves at the level of syntax; otherwise the predicates that occur in the two sentences and the relationship between the predicates should count. But Russell is not paying any attention to these matters.

[8] Russell (3), Chapter III, p.69; whatever may be way in which 'significant proposition' as it is used here is understood, it will (have to) include that it is an expression which is either true or false.

[9] See below the section entitled "Some Indispensable Exegesis."

[10] Russell (3); emphasis added.

[11] ibid; emphasis added, and notational changes made for convenience and uniformity. What is the definiendum and what is the definiens here may be noted with care.

[12] Existence and predication are independent of each other whether one takes the former in the narrow sense like Russell, or in the extended sense identifying it with 'Being' as Meinong did. Predication does not produce entities, nor do entities require predication to come into being unless they are beyond the threshold of acquaintance in the Russellian sense of the term. We are given "qualitative particulars," I trust.

[13] This however does not constitute a test for deciding whether χ is an instantiated predicate or not; this is determined by its presupposition.

[14] See below the section entitled "Some Indispensable Exegesis."

[15] This may be seen in the context of Russell's skepticism about the significance of Frege's way of providing nominata to non-denoting terms, namely his way of assigning the null set as their common denotatum; for Russell such strategies are unnatural and hence devoid of significance.

[16] Evidently a sentence with a definite description as its subject term can be false either by failing on the count of uniqueness or by failing on the count of existence.

[17] See footnote 40 below.

[18] To scan through the vast literature in this area to note how Russell was understood on a specific point is like locating a needle in a haystack on the Moon. So I focus my attention on Strawson (1), Linsky (1), and Consuegra (1).

[19] There is, of course, more to these than just that; the first involves "diversity", and the second involves "identity." The truth of the first needs a two membered universe, and the truth of the latter requires only a universe with at least one member. Russell is quite aware of this difference; see Sec.50 of his (3) "Identity and Diversity as Relations.' I shall not go into details concerning this point for it is beyond the scope of my concern here; and my comment on this issue is available in Part Two of this volume. It may be recalled that my way of formulating the presupposition and entailment of a negative sentence with a definite description as its subject term encapsulates just this much; it has the additional virtue that it is formulated as a syntactical principle. I may not have completely succeeded in my attempt; but success on that count is the sine qua non for totally redeeming negation from ontological affiliations.

[20] Emphasis added in the third and the fourth cases.

[21] See Chapter Two above.

[22] I am not framing these issues the way Strawson or Linksy framed them; and I am induced to take to this way of putting them by Russell's reply to Strawson. Strawson's point was that Russell held that a part of what we affirm when we affirm the definiendum is in fact the definiens, and that he could hold this because he failed to note that (a part of) the definiendum is presupposed by the definiendum. Russell's reply, or his defense of what he did, was: "My theory of descriptions was never intended as an analysis of the state of mind of those who utter sentences containing descriptions;" it was, on the other hand, intended "to find a more accurate and analyzed thought to replace somewhat confused thoughts which most people at most times have in their heads." See his (11).

[23] Russell (11); emphasis added. For once Russell uses 'sentence' instead of his habitual 'proposition'.

[24] Russell (I).

[25] p.174; emphasis added. See endnote 2, Chapter One. Here Strawson is using Rusell (2) while discussing Russell (3).

[26] (1), p.122; his 1987a, his doctoral dissertation i.e., has been published since then as his (3).

[27] I wonder how many of those unpublished writings can be credited with relevance, all the more so when Consuegra himself is convinced that, to quote him, "one can prove almost anything from them;" see his (2). This is why our curiosity does not rest when we learn from Consuegra that Russell, in his unpublished writings, specially his "On the Meaning and Denotation of Phrases" (25 folios), subscribed to what Strawson held in 1950. The fact of the matter is that these unpublished compositions were written before 1905, and not either in or after 1905. This means that Russell considered the view in question, found it wrong, abandoned it, and developed what he thought was the right view in 1905; his not rushing to the printer with those compositions in hand is indicative of this. Consuegra does not tell us either in his (2) or in (3) why Russell changes his views on this point. In any case the reasons for this change can be elicited from Russell's published writings. Thus knowledge of the unpublished writings can only be likened to the ability to answer the questions of quiz-master. It is so pointless that credit should go really not to those who assiduously scan them, but to those who ignore them, or else to those who do not have access to them. Odd details, in order to be interesting or significant must really be odd, otherwise they do not enlighten - they do not amuse either, they only bore us.

[28] This means that, as it can easily be noticed, to exist is to be a value of a bound variables; Quine, of course, is the first to shout: Ahoy ! Land.

[29] Russell (3), p.174.

[30] When $(\mathtt{I}x)\phi x$ does not exist, $(\mathtt{I}x)\phi x = (\mathtt{P})\mathtt{P}$ where $\mathtt{P}$ is an apparent variable ranging over propositions, as Russell is reported to have remarked in some unpublished work referred to by Consuegra; the idea here is the same as the one which Post brilliantly adumbrated and which culminated in his classic completeness theorem.

[31] Despite temptation, I resist being dragged into the metaphysical mire; so I shall be brief, and move ahead with remarks which can be elaborated eternally.

[32] Geach (1), 3rd edition, 1980. Geach's remark about Oxford philosophers seems to be more appropriate in the case of free logicians, as we shall see in Chapter Five. Emphasis is added.

[33] Frege (4), p.97; emphasis added.

[34] This has been the source of considerable irritation to Wittgenstein; see Rao (5).

[35] Wittgenstein (1), and Rao (5).

[36] Wittgenstein (4).

[37] Nevertheless Russell mentions such predicates, specially in (3), but he does it in the contexts where he is arguing that even if they are inducted into our armory, problems concerning identity, and the categorical difference between proper names and definite descriptions, still remain unresolved.

[38] This has bearing on the extensionality thesis, and the substitutivity of identicals. I will return to this in the next chapter when I consider Dana Scott's criticism of Russell's theory.

[39] Russell continued to hold this view even after his (3). Wittgenstein, recapitulating his interactions with Russell on this issue during 1914-1916, records in 1937:

> "In the course of our conversations Russell would exclaim "logic's hell" - and this perfectly expresses the feeling we had when we were thinking about the problems of logic; this to say, their immense difficulty, their slippery texture."
> "I believe our main reasons for feeling like this was the following fact: that every time some new linguistic phenomenon occurred to us, it could retrospectively show that our previous explanations were unworkable."

See Wittgenstein (6).

[40] Russell (4), Chapter XVI.

[41] Wittgenstein articulated this idea in his (1); however he thought that in a syntactically perfect language, each expression generates its own identity.

[42] This is precisely what is encoded in the rule of Existential Instantiation; a constant associated with a variable in applying this rule will have to be kept associated with it throughout the argument (discourse).

[43] We may call these respectively existence, constancy, and uniqueness assumptions.

[44] Frege (3); emphasis added. It may be recalled that Frege includes definite descriptions too in the category of names.

[45] Russell (2); note how the distinction between propositions and facts gets blurred here. See Rao (6).

[46] The notion of a quantifier, as Russell would say in (8), p.73, is "(a)n infinitely complex concept ... (and) ... can certainly not be manipulated by human intelligence." Nevertheless, its nature can be explained by the way it is used, or variantly in terms of the rules of inference. It is essentially a device by means of which we represent the infinite in finite rubrics. That a generalized sentence is either a logical product or a logical sum is the idea of the algebraists from Skolem onwards. Incidentally, Ramsey was in the tradition of the algebraists; see Ramsey (1).

[47] Grattan-Guinness (1). Russell's 1906 manuscript "On the Substitutional Theory of Classes and Relations" on which he based his judgment has since then been published in Russell (14). See Hylton (1), and Landini (1).

[48] We may differentiate these sentences form those of the form ~(Ex)ϕx by calling the latter "negative particular" sentences.

[49] A sentence can be taken as false if it is not true; on this reading truth-value gaps disappear.Such a reading of 'is false' even for atomic sentences was suggested by Kaplan in his (1). Given the fact that the range of the variable x in both (Ex)ϕx and ~(Ex)ϕx is the same, if the former is false the latter would be true. The general spirit of Russell's theory is analogous to his definition of falsity. Just as he first analyzes 'truth' in terms of correspondence, and then defines falsity in terms of truth, Russell first explains how phrases that do indicate succeed in doing so, and then in terms of this explains failures of indication.

Chapter Four : Crossing 'T's and Dotting 'I's in Russell's Draft

In this chapter I shall take into account some important objections to and criticism of Russell's theory, and try to show how they are based on serious misunderstandings of Russell's program, and how they can be met within the framework of Russell's theory. My intention thus is far from completing Russell's program whatever such completion amounts to. The silliest thing a philosopher can do is either to debunk another philosopher or else to vindicate his contribution. As philosophy is not apologia, what I propose to do in this chapter should be nothing more than a specification of some reasons for my preferring Russell's theory over others. The objections to and the criticism of Russell's theory that I propose to take into account are from David Kaplan, Dana Scott, and Abraham Robinson[1]. In considering the criticism of these logicians and philosophers to show how they do not adversely affect the plausibility of Russell's theory of definite descriptions, or why it should be preferred over the alternatives suggested either by them or others, I shall observe two norms of criticism which should be adhered to by everyone including those critics. The first of these is to attack the opponent at his strongest point; intellectual fairness requires this. The second is not to shift the point of reference in the debate. It hardly needs to be emphasized that within the framework of a philosophical debate in which a problem unfolds, something always seems to be decisive to everyone, and from that point of view the opponent seems to have been routed. But more often the framework itself may be doubtful, and when it is doubted all the aura of solid achievement starts disappearing and the presumed success of various arguments and strategies adopted become problematic. The strength of a philosophical offense thus depends upon concentrating forces on a very narrow point as much as the strength of philosophical defense depends upon keeping the architectonic secure. Having thus indicated my strategy in countering the critics of Russell, I may first consider Scott's criticism of Russell where he shifts the problem and bypasses Russell's strong points while being fully aware of them. Scott is unhappy with Russell's theory on two counts:

> "In many ways Russell's idea is the most attractive because of its simplicity. However," he writes, opening his argument, "on second thought one is saddened to find that Russellian method of elimination depends heavily on the scope of elimination. Further, the semantical meaning of Russell's transformation is not all that clear although it could be made precise."[2]

This means that Scott would have preferred Russell's theory but for the fact that, as he notices, the complexity of the scope mechanism adversely affects its simplicity; and secondly it means that his own alternative to Russell's theory brings out the "semantic meaning"[3] of definite descriptions while scoring better than Russell on the count of simplicity. Testing this claim partly hinges upon the meaning of "simplicity." But contemporary mind is trained to understand the complex and not the simple; consequently it seems that it cannot understand what simplicity is. So a word about complex notion of simplicity itself. We may try to specify the meaning of simplicity in a way that would not only be acceptable to Scott, but would indeed be advocated by him, as he is one of the foremost logicians today. For the simplicity in question is not epistemological but logical, it should be pleasing to Scott that what is simple is, for instance, amenable to straightforward (mathematical) induction. Unless Scott is taking simplicity in some non-logical sense, if not specifically epistemological sense - in which case we may ignore his criticism - his objection to Russell's theory will have been met to his satisfaction if it can be shown that the occurrences of multiple definite descriptions in a sentence, and their figuring in it one in the scope of the other, are amenable to inductive procedures, and hence do not involve complexity in the logico-mathematical sense of the term. There is nothing ingenious in this. Scott himself could have noticed this if only he cared to do so; and Russell himself could have spelled it out had he thought that it was not obvious or was beyond the threshold of the working logician's normal perception. In any case Rusza[4] has given us one way of doing this, and we may note in brief an alternative way of doing the same. We may begin with something to which Russell paid considerable attention in his (2), namely the occurrence of a definite description in a sentence being either primary or secondary. The relationship between this conceptual pair and the notion of scope are intimate, and intricate as well, leading to some glaring if not gross misunderstandings. First, the distinction between the primary and the secondary occurrence of a definite description in a sentence is final and absolute, according to Russell obviously. This is to say that the conceptual pair in question is exclusive and exhaustive; there is no third way in which it can occur in a sentence[5]. Further, the theory of descriptions is primarily concerned with primary occurrences of definite descriptions; of course, Russell spelled out a strategy for deploying the results arrived at in connection with these to handle the riddles that crop when definite descriptions occur secondarily, specially puzzles that arise on the count of their occurrences in epistemic and intentional contexts. Russell's point was that the reduction sentence $R\{\chi(Ix)\phi x\}$ of $\chi(Ix)\phi x$ invariably takes the shape of an existentially affirmative, this is to say that it will assume the form $(Ex)(\ldots)$, if the occurrence of $(Ix)\phi x$ in $\chi(Ix)\phi x$ is primary. The problem thus is: how to proceed with the reduction when several definite descriptions occur in a sentence, and occur in the primary sense of the term. Thus the problem is more of practical nature than theoretical. We can think of the first reduction, the second reduction, the third reduction and son on, and not - as Kaplan thought that it should be - of primary, secondary, tertiary, ..., occurrences of a definite description occurring in a sentence; and such an ordering is only a preparation and a prerequisite for the projected reduction, and is not a part of it.

Given that $(\mathrm{I}x)\phi x$ is the canonical abbreviation for "the term x which satisfies ϕx"[6] and that a primary occurrence of $(\mathrm{I}x)\phi x$ in a sentence involves the existential quantifier, that $(\mathrm{I}x)\phi x$ exists (this is to say that the x satisfying $\phi!\hat{x}$ exists, which may be shortened as $E!(\mathrm{I}x)\phi x$) may be mapped onto $(Ex)\{\phi x.(x)(\phi y \equiv x = y)\}$. We may call this the "existence sentence" of $(\mathrm{I}x)\phi x$. An occurrence of a definite description in a sentence is to be considered as <u>outermost</u> if it does not lie inside another definite description. For all sentences ϕ, we may define the first existence sentence of ϕ - ϕ^1 for short - by induction, as follows:

(1) If ϕ is atomic, and no definite description occurs in ϕ, then $\phi^1 = E!^0$; this is to say that the existence sentence for this case is $(Ex)(x = x)$.

(2) If ϕ is atomic, and t_i, ..., t_n are all the outermost descriptions of ϕ, than ϕ^1 is the conjunction of the existence sentences of these definite descriptions; thus

$$\phi^1 = E!t_i., \ldots, .E!t_n.$$

To put it differently, as the sentential connectives here are classical, the existence sentence for this case will be $\sim(E!t_i \rightarrow, \ldots, \rightarrow \sim E!t_n)$.

(3) $(\sim\phi^1) = \phi^1$.

(4) $(\phi \rightarrow \psi)^1 = \phi^1$ if $\phi^1 = E!^0$;

otherwise

$$(\phi \rightarrow \psi)^1 = \{\phi^1 \text{ if } \phi^1 = E!^0, \text{ if not } \{\phi^1 . \psi^1$$

(5) $\{(x)\phi\}^1 = \{(Ex)\phi\}^1 = \phi^1$ if ϕ^1 involves no free concurrence of x, otherwise

$$\{(x)\phi\}^1 = \{(Ex)\phi\}^1 = (Ex)(\phi^1).$$

Obviously, if ϕ does not contain a definite description, $\phi^1 = \mathtt{E!}^0$ as noted earlier.

The existence sentence of $(\mathtt{I}x)\phi x$ does not define $(\mathtt{I}x)\phi x$, in the strict sense of the term. What Russell tries to define in 14.01 of the Principia is not $(\mathtt{I}x)\phi x$ as such, but its occurrence in a sentence, how it behaves in a sentential context, that is. There is an unbridgeable gulf, semantic gulf, between the existence sentence onto which a definite description is mapped, and the definiens of a definition in which the definitum is a sentence containing that definite description; $(\mathtt{I}x)\phi x$ is only a definiendum which is a part of the definitum. In a contextual definition, the definitum and the definiendum are not the same, and in fact belong to two different syntactical categories. It is the definiens of a contextual definition which is a reduction sentence. The existence sentence of the definiendum, and the reduction sentence of the definitum may not have the same truth-value; the former is only a semantic part of the latter. The semantic difference may as well be represented syntactically by ranking sentences, and by showing how the rank of the reduction sentence corresponding to an occurrence of $(\mathtt{I}x)\phi x$ is always higher than the rank of its existence sentence. We may define the rank of a sentence as follows:

(1) if t is an individual variable, then the rank of t - $r(t)$ for short - is 0.

(2) $r(t_1 = t_2) = r(t_1) + r(t_2)$.

(3) if P is an n-place predicate, $r(Pt_1,\ldots,t_n) = r(t_1) + \ldots + r(t_n)$.

(4) if ψ is either $\sim\phi$, or $(x)\phi x$, or $(Ex)\phi x$, then $r(\psi) = r(\phi) + 1$.

Finally

(5) $r(\phi \rightarrow \psi) = r(\phi) + r(\psi) + 1$.

As it turns out if the rank of ϕ is n, the existence formula of $(\mathtt{I}x)\phi x$ is of the rank $2n + 7$; and the rank of the reduction sentence corresponding to the sentence containing an occurrence of $(\mathtt{I}x)\phi x$ will be greater than $2n + 7$. In general the rank of ϕ^n will be lower than the rank of ϕ^{n+1}, unless $n = 0$. All this is not intended to show how syntax can be arithmeticised; it is geared, on the other hand, to indicate how scope confusions can be overcome by simple procedures. Given a sentence, say $\chi(\mathrm{I}x)\phi x$, we can proceed to produce the desired reduction sentence first by noting the scope of the existence of $(\mathrm{I}x)\phi x$, and then applying the definition 14.01 of the Principia on the sentence in which $(\mathrm{I}x)\phi x$ occurs whatever be its complexity or the complexity of ϕx.

Thus, in so far as logical simplicity can be understood as amenability to inductive procedures, Russell's theory cannot be found faulty on the count of simplicity. Moreover, complex explanations of simple phenomenon are as desirable or undesirable as are simple explanations of complex phenomenon. A paradigm explanation, whether plausible or not, should strive towards that degree of complexity or simplicity which is commensurate with the nature of the phenomenon; and on that count Russell's theory scores very high, at least higher than the alternatives suggested by Scott, as we shall show soon. Further, there seems to be some truth in what Hintikka says in his discussion that follows Kaplan (I), namely that the scope conventions are not conventional, and that they are chosen differently on different occasions.

> "There are," he says, "even good reasons for assuming that no scope conventions can do justice to ordinary language;"

so we need to settle for marginal sins, as Russell seems to have realized.

The other fault which Scott notices in Russell's theory is that the semantic meaning of the Russell-transformation is not clear. The transformation at issue must be the one Russell suggested in 14.01 of the Principia, namely

$$\{(\mathrm{I}x)\phi x\}\chi(\mathrm{I}x)\phi x =_{\mathrm{Df}} (\mathrm{E}x)\{(y)(\phi y \equiv x = y) . \chi x\}$$

and not the one which was encoded in 14.02 which we considered as a mapping of $(\mathrm{I}x)\phi x$ onto $(\mathrm{E}x)\{(y)(\phi y \equiv x = y)\}$, and is only auxiliary importance. If there is any unclarity in a Russell-transformation it could be either due to the opaqueness of the semantics of the system to which the definiens belongs, or else due to the vagueness in the equivalence of the definitum and the definiens of 14.01. Scott could not have meant the former, for that would amount to suspecting the standard quantification theory and its standard interpretation; this Scott does not do. Thus he must have had the latter on his mind. Then, in view of what emerged from the clarification in Chapters Two and Three earlier, Scott's contention stands without any legitimacy and credibility. If our goal is to capture the basic semantic aspect of language - either natural language in its received grammar, or natural language with some reformed version of it - then elimination of truth-valueless sentences from the realm of semantics should be our overriding concern. Failure on this front amounts to failure to capture something very essential to the functioning of language as a toll of representation. Recognizing that Frege[7] failed on this count, Russell tried to overcome the difficulty by fixing the truth-value, and thus the meaning (for him a sentence without a truth-value is devoid of meaning, and is no sentence at all in the strict sense of the term) of a sentence by assigning it a syntactical place in the network of sentences that constitute the language. Using the Wittgensteinian lingo we may as well say that Russell tried to assign a place to such an expression in the syntactical space of the language. The semantics of the unproblematic sentences is shown to envelop, in virtue of their syntactical relationships, even semantics of the initially problematic sentences. Thus what Russell took to be semantically unproblematic sentences are problematic to Scott, it seems.

In order to secure for the Russell-transformation what, according to Scott, eludes it in Russell's way of handling it, in the alternative he worked out, Scott married it off to what Russell rejected, namely Fregean semantics; however, he did that pushing aside the conceptions of language, logic, and their relationship which were shared by Frege and Russell. This means that he has shifted the very area of debate. What then Scott means when he says that the semantical meaning of Russell-transformation was not clear is that it is not clear that which model Russell had on his mind. The fact of the matter is that Russell did not have even the notion of model on his mind, thus in a pertinent sense Scott's objections to Russell's theory are off the mark; they are tangential, at least. Hence, Scott's contention that his alternative brings precision to Russell's unclear intentions is not only misleading but is a distortion of those intentions as well. Before considering Scott's alternative and its conceptual apparatus we may note that his choice, as he states it with admirable clarity, was motivated - that is his term - by three principles which are: (1) that the bound variable should - that again is his term - range over the given domain of individuals, (2) that the domain of individuals should be allowed to be empty, and (3) that the values of terms and free variables need not belong to the domain of individuals. The last principle, in its turn, was motivated by its utility - in the pedestrian sense of the term. Though all this is conceptually alien to Russell's thinking, we may translate his point of view into this lingo to note that Russell would not have subscribed to any of these three principles; in fact he would have outright rejected each of them. In his frame of things, the range of free and bound variables is precisely the same; the quantifier only sets apart those instances of the substitution of free variables for which the resultant sentences are true. And in virtue of the fact that in the canonical language of the Principia there are no uninstantiated predicates, the second principle would have been treated by Russell as patent falsehood. More importantly, the issue is not what should or need to be the world so that our language might represent it, but what is the way in which language and what it does represent - the world, that is - are related together. The issue of choosing an interpretation among the possible one does not arise at all. Interpretation is internal to languages; and logic is descriptive of the unique way in which a language succeeds in representing the world - "this one actual world!" Russell's conception of language is not legislative, it is bureaucratic; only, he wanted to infuse into it a little more efficiency by streamlining its grammar to make it transparent. Language as well as the world it represents - its model, if one likes to put it that way - both are given to the logician; he invents neither of them. Language is given, and what is left to the logician is to make its grammar transparent, precise, and systematic. On the other hand, for Scott language is a constructable symbolic system; thus in constructing it practical needs may, and can, be taken into consideration. (This is a marriage, harmonious or otherwise, of subjectivism and pragmatism from which issues all that is central to and explicit in the American way of life and thought!) As Russell and Scott are poles apart in their respective conceptions of language and logic, it cannot be said that Scott's system is an alternative to Russell's; nor can it be held that they are solving the same problem. Russell belonged to what Hintikka calls the Universalist tradition of language, which is called that as it upholds that language is a universal medium, and Scott belongs to the Model Theoretic tradition, or with a less picturesque name, to purloin another of Hintikka's phrases, the tradition of the possible-world semantics, according to which language is a calculus. As possible models are not indiscernible, and as language is about each of the models, with reference to each of them it finds a distinct interpretation. There are too many metaphysical mermaids in the waters of possible worlds, but I will not be interested in them. What concerns me here is that all this is alien to Russell's view of the matter. According to the universalist conception of language, it does not have alternative interpretations, nor are reinterpretations of it possible. Thus the formalized system which Russell projected in the Principia, like the Begriffsschrift of Frege, is an interpreted system. The conception of logic which goes with it is different from the conception which is embedded in the model theoretic approach to language. As

> "(t)here is no sense in talking about other possible worlds or even other models (interpretations) of our language ... logical truths," as Hintikka noted[9], "cannot be explained a la Leibnitz (or Carnap) as truths in every

> possible world. Logical truths are for Frege truths about this world of ours, albeit about its more abstract aspects. This explains Frege's characterization of analytic truths as being based exclusively on completely general laws and definitions ... These characterizations are based on the assumption which Russell once expressed in his inimitable way of saying that "logic is concerned with the real world just as zoology, though with its more abstract features," ... These views were in the main shared by early Russell, Wittgenstein ... I would not even be surprised if a perceptive historian of twentieth century thought could trace the same source to Chomsky's emphasis of syntax at the expense of semantics."[10]

The thrust towards finding a solution to the problem of definite descriptions using only syntactical apparatus is all too evident in Russell's writings to be news to anyone, and so are the reasons for his attempting to have such a solution. He is explicit about it in his (3) where he writes:

> " ... a definition is strictly speaking no part of the subject in which it occurs. For a definition is concerned wholly with the symbols, not with what they symbolize. Moreover it is true or false being the expression of a volition, not a proposition."[11]

This should exhibit without any unclarity how disparate are the volitions of Russell and Scott, and how the latter cannot be, in any legitimate sense of the term, said to be a precise version of the former.

We may now turn our attention to all those virtues which Scott claims towards the superiority of his version over that of Russell. The most important of these is that it is "quite natural," and that it "corresponds quite faithfully to ... natural modes of reasoning." Scott's claim may be justifiable when his alternative is juxtaposed with Hilbert's strategy, but not with Russell's because what he invoked to make that claim is what he, in fact, borrowed from Russell, namely that it is natural to use definite descriptions before it was ascertained whether they are proper ones. Of course, the reason he gives, namely that otherwise the class of sentences will be deprived of its recursive character cannot be credited to Russell according to whom the class in question is given, and not generated. But with his semantic interpretation Scott loses more than he gains on other counts, and loses all his gains by "insisting that improper descriptions all assume the improper value;" and he does insist on this to have the "highly useful law of extensionality $\vdash (x)(\phi x \rightarrow \psi x) \rightarrow \{(\mathrm{I}x)\phi x = (\mathrm{I}x)\psi x\}$." Obviously, this would not be valid

> "If one wanted 'the golden mountain' and 'the round square' to have different values;"

and Scott continues,

> "while making unkind remarks about 'the golden mountain' Russell also rejected this law of extensionality which this author (i.e. Scott) considers an unfortunate choice."

Russell's "choice" however is not at all unfortunate as Scott sounds it to be; it is on the other hand well thought out and was intended precisely to be natural - and commonsensical too - for 'The golden mountain is (a) round square' is not a trivial identity like a = a [12], nor are 'The golden mountain is round' and 'The round square is a mountain' about the same in any sense which is natural and commonsensical.

> "It is an old dispute," writes Russell in (3), "whether formal logic should concern itself mainly with intensions or with extensions. In general, logicians whose training was mainly philosophical have decided for intensions, while those whose training was mainly mathematical have decided for extensions. The fact seems to be that, while mathematical logic requires extensions, philosophical logic refuses to supply anything except intensions. Our theory ... recognizes and reconciles these two apparently opposite facts, by showing that an extension ... is an incomplete symbol, whose use always acquires its meaning through reference to intensions."[13]

The model-theoretic approach dispenses with the overriding concern with this world - a concern which is all pervasive in Russell - in understanding what constitutes a logical truth. Logical truth, according to model theorists, is only a matter of language itself. If Quine is the last survivor of the Russellian tradition, Hilbert is the archpriest of the view that logical truth is an internal affair of language. This may appear to be a misrepresentation of the model theoretic view because the notion of a model, or of possible world, looms large in that kind of characterizing what constitutes a logical truth. But if we pay some attention to the conceptual apparatus involved in it, and resist being bewitched by the seemingly high-powered results of model theory, we will have noted the non-natural notion of language, indeed an unnatural notion of language, embedded in the model theoretic approach. According to this view of language, a language is a set S of sentences, closed under a finite set of rules of formation and transformation and having a model[14]. As I will not be able to improve over the great votary of this cult, namely Hintikka, I may as well quote him.

> "According to this ... model-theoretic conception, what a sentence S says, it says by specifying a class of models M(S). To know that S is true in a model M_0 is to know that $M_0 \in M(S)$. To know that S is logically true is to know that it is true in every model, i.e. to know that M(S) = the entire space of models. Thus in the last analysis everything comes down to the relation of a sentence S (or a set of sentences) to its models that the non-logical constants may be reinterpreted ... What the elements of (the model) are does not matter in the least ... They may be elements in some abstract Platonic structures; they may be interpreted (as) elements of another kind ... None of these ideas make any general sense to someone who seriously believes in the universality of (language)."[15]

If these ideas do not make sense to someone, it could be due to the fact that one who subscribes to the universality of language is not endowed or equipped with the skill or brilliance required to appreciate the model theoretic battery, or else his finding that this battery is devoid of all those virtues that are claimed to its credit. Russell seems to have stuck to his position not due to the former reason; though his terminology and apparatus are different from that of the model theorists, he was indeed tackling the central issue of their "new-think." That he upheld the universalist position due to the latter reason can be entertained because in a sense, perhaps in a charitable sense, Meinong was reinterpreting our language. Meinong's interpretation, in a pertinent sense, defines a model which falls part of the logical space of the models of our language as a model-theorist envisions it. Russell had reasons for rejecting Meinong's reinterpretation as any reader of Russell (2) and (3) should know. The issue, to speak Model theoretic lingo, is whether the members of the domain of an arbitrary model M_i such that $M_i \in M(S)$ bear any resemblance to the objects which form the "substance of the world," to put it in the terminology of the Tractatus. Even if we do not frame the problem in this way, it should be clear enough. Unless this resemblance, if not identity, is established, the calculus C with an interpretation such that its model M_i is a member of $M(S)$ remains inutile as a mechanism of representing our world. It hardly needs to be highlighted that this is not just another issue of mapping that language C (this is to say that, the calculus C with the interpretation in question) onto our language, which we may refer to as L, with its received interpretation. It is much more than that. It is about establishing the intensional linkage between C and L. (Pace the reference to intension in the passage from Russell (3) quoted above.) Despite Hintikka's defense of Hilbert, or for that matter, Hilbert's befuddling of Frege, Frege still had a "philosophical point" concerning Hilbert's axiomatisation (of geometry.)[16] A non-Euclidean geometry may be mapped onto Euclidean geometry[17], but this itself is not sufficient for legitimately believing that the geometry so mapped does represent the structure of our perceptual space which still remains (to us) Euclidean as our earthly Euclidean minds have us believe. This does not mean, as Hintikka would have us believe, that we are prisoners of our language, for all that it indicates is that we are not free from what we are. Thus until the tangible connections between alternative possible interpretations on the one hand, and the received interpretation of our language on the other are established to the satisfaction of the proficient speakers amongst us, it does not follow that our language under a reinterpretation does represent the world as it is given to us - or, represent our world, if one is too possessive. These tangible connections cannot be established in terms of truth, for the model theoretic notion of truth is relative; it is relative to the model with reference to which it is defined. It is evident that model theoretic definitions of truth are geared to recursively define 'true in M'. This naturally leads to far too many problems which have been much debated about in the recent logico-philosophical literature; but there is no need to recount them here. The identity of interpretations that we were talking about is the business of definitions; what Russell claims for his definitions, specially 14.01 and 14.02, is not that they are true, but that they are plausible. (Pace the concluding lines of his (2) where he suggests a link between the intensional and extensional realms.) This is why such definitions do not form part of either of the interpretations. To repeat,

> " ... a definition is strictly speaking no part of the subject in which it occurs."[18]

Russell's categorization of 14.01 and 14.02 of his (3) itself has been called into question by Kaplan who extended some of the ideas of Scott. As his argument is intended to show that Russell was unclear on, or unsure about, many crucial points concerning definite descriptions, it seems necessary to pay some attention to his argument and show that it is based on a misunderstanding of Russell; only after this, can we return to Scott to show how his claims about his alternative to Russell's theory cannot be sustained. Kaplan's central argument is that 14.01 is not a definition because $(\mathrm{I}x)\phi x$ has, according to Russell, no meaning in isolation, and hence the right hand side of 14.01 cannot be said to be either explicating or else bringing precision to the meaning of it or even $\chi(\mathrm{I}x)\phi x$. This is contrary to what Russell intended it to be. That apart, it does not have a meaning in isolation, and that is precisely why, as Russell argued, it had to be defined in the context of its occurrence in say $\chi(\mathrm{I}x)\phi x$. Now, to suggest that $\chi(\mathrm{I}x)\phi x$ does not have meaning because $(\mathrm{I}x)\phi x$ does not have meaning in isolation, is to assume the universal validity of the Fregean Constitution Principle in a strengthened version. This assumption however can easily be shown to be in effect equivalent to the assumption that there exists a unique correlation between rules of formation and rules of evaluation. As Kaplan would say:

> "This has the desired result that the semantical evaluation of an expression exactly recapitulates its grammatical construction."[19]

Russell's point was <u>not</u> that $\chi(\mathrm{I}x)\phi x$ lacks meaning (sense, connotation) when $(\mathrm{I}x)\phi x$ is devoid of an indication; it is obvious that he assigned a truth-value to $\chi(\mathrm{I}x)\phi x$ when $(\mathrm{I}x)\phi x$ is improper. On the other hand, it was that $\chi(\mathrm{I}x)\phi x$ has a meaning, thus a truth-value, even if $(\mathrm{I}x)\phi x$ is devoid of an indication. What he was trying to do was to formulate a sentence which contains the same non-logical symbols as $\chi(\mathrm{I}x)\phi x$, and which first has the same truth-value as this sentence, and secondly whose truth-value is determined by the indication of expressions that figure in it. Thus in the Russellian framework, the Fregean constitution principle holds without any proviso so far as the realm of meaning is concerned. This is so because the truth-value of a sentence is not, for Russell as it was for Frege, its detonatum (referent, nominatum.) Thus there are, in a sense, two constitution principles, one holding for expressions figuring in the definitum, and the other holding for expression occurring in the definiens. That all and only grammatically well-formed expressions with sense (this is to say that, sentences) have a truth-value links these two principles and thereby the realms of intension and extension. Perhaps, Russell surmised that a non-gap evaluation is the <u>sine qua non</u> for projecting a plausible connection between extension and intension and hence stuck to it.

In view of this it can be said that Kaplan's rhetoric,

> "(b)ut what is the meaning given to it, or even the full description? None! For the central thesis of Russell's theory is that this phrase has no meaning in isolation,"

misses the crucial point of Russell which is that even if $(\text{I}x)\phi x$ does not denote, $\chi(\text{I}x)\phi x$ has a meaning, and hence a truth-value which is the same as the value of the sentence which figures as the definiens in its definition on the pattern of 14.01. However, this equivalence is not strong enough to be considered as identity, and hence the definitum and the definiens in 14.01 do not have all their properties in common. For instance, the notion of denotation may be totally irrelevant in the case of the definitum, and the fate of the definiens may exclusively hang upon this notion. Kaplan has not either cared to notice, or else has ignored this stand of Russell. That Russell understands 14.01 and 14.02 of the Principia in the way I understand them here should be evident form what he says in his (3). I am not foisting an interpretation on it, nor am I reading into it. In support of my view of things I may just juxtapose 14.01 and 14.02 with the opening lines of Chapter 13 of (3). Explaining his definition of identity, namely

$$x = y =: (\phi).\phi!x.\rightarrow.\phi!y \text{ Df.},$$

Russell remarks that the

> "use of the sign of equality (in the above i.e.) covers all the common uses of equality that occur in mathematics,"

and cautions:

> "Note that the second sign of equality in the above definition is combined with Df; and thus is not really the same symbol as the sign of equality which is defined. Thus the definition is not circular, although at first sight it appears so."

Thus Kaplan's objection that 14.01 cannot be treated as a definition because in it the definiendum is improper is not well-founded. And though he rightly thinks that $(\iota x)\phi x$ cannot be treated as an abbreviation of $(Ex)[\{(x)\phi x \equiv x = y\} . \phi y]$, he does it for wrong reasons which need not bother us here as our concern is not Kaplan's confusions but with his misunderstanding of Russell's position. Kaplan's preferred understanding of 14.01 however is that it is a rule for translating the logically imperfect language into a logically perfect symbolism; but he has a lingering doubt which he formulates as follows:

> "if Russell really took his doctrine seriously and were willing to completely abjure the 'misleading' surface grammar of definite descriptions, he should be willing to accept the present understanding of (14.01). His use of definite descriptions in Principia Mathematica indicates to me a linguistic ambivalence."[20]

From Russell's remark that a definition is no part of the subject in which it occurs, it should follow that he intended 14.01 and 14.02 to link two realms, not two structures, for this term has model-theoretic connotations which are alien to his thought. That these two realms are not two different languages should also follow from the fact that he subscribed to the universal view of language. This means that 14.01 cannot be, in the strict sense of the term, a rule of translation; it can thus be a rule of transformation, at most. But then what is that which it is intended to transform, and what is it that it is geared to transform into? Given that the definitum or the definiendum, and the definiens of 14.01 are expressions of the same language, and given that $(\iota x)\phi x$ is a transparent representation or a precise reformulation of the opaque so-and-so which is such-and-such, it is difficult to see why the transformation projected in 14.01 was thought of at all. $\chi(\iota x)\phi x$ is not devoid of meaning, for we do understand it as well as we do; and we understand it even if $(\iota x)\phi x$ is improper. This is to say that we know what it after all means; only we run into troubles when we try to ascertain whether it is true or false, when we start attempting to fix its truth-value, that is. If the notion of truth were to be understood in terms of denotational semantics, the (desired) parallelism between rules of evaluation and the rules of formation will break down in the case of expressions of this form, or will lead either to unnatural entities or to truth-value gaps. Speaking anachronistically, Russell's answer to Kaplan's rhetorical query:

> "(d)oes acceptance of the equivalence (14.01) commit us to Russell's analysis of the logical form of sentences containing definite descriptions?"[21]

is implicit here, and it is, contra Kaplan, a firm 'Yes'. The alternatives that Kaplan takes note of, namely the Frege-Strawson and the Frege-Carnap approaches, are to be kept out, and expressions of the form $\chi(\mathtt{I}x)\phi x$ will have to be treated as logically imperfect in the Kaplanian sense of the term, or as grammatically deceptive in the Russellian sense of the term., Their grammar does not tell us, to put it in Kaplan's words

> "how to 'construct' the semantical value of an expression in terms of the values of its logically simple components."

Kaplan wants this semantic evaluation be a one-stroke affair; this would enable us to keep it on a recursive base. He includes such an evaluation in the defining features of a logically perfect language. Thus so far there is nothing wrong with his argument. But it goes astray when Kaplan starts thinking that Russell's canonical language in (3) to which the definiens of 14.01 belongs is logically imperfect because it fails on the count of one-stage semantic evaluation, and that it fails because expressions like $(\mathtt{I}x)\phi x$ are terms of that language. However, this should not surprise anyone as it is our infirmity that we very often miss the obvious. Conflation of two notions of equivalence embedded respectively in the use of the sign = in 14.02, and its use say in 14.20 - variantly put, the use of = in 14.02 and the use of ≡ in 14.20 - and his belief that $(\mathtt{I}x)\phi x$ is a term of the canonical language of (3) go hand in hand in Kaplan's argument; one leads to and reinforces the other. This conflation and this belief resulted in Kaplan's noticing a "lingering ambivalence" in Russell and thus coming to hold that Russell could not make up his mind whether (Ix) is a quantifier or a term producing device. It is not the case that the Principia is free from ambivalences; we shall point to some a little later, but they are of a different sort. However, Russell is not Kaplan-ambivalent. He was clear, as we pointed out in Chapter One, that (Ix) is not a quantifier; nor is it a term producing operator in the language of the Principia. Kaplan's argument that Russell intended to extend the vocabulary of that language does not hold good, for his intention was to eliminate all expressions whose logical or orthographic characterization requires them. To subscribe to what Kaplan is suggesting amounts to holding that Russell introduced a class of terms and then eliminated them, like the proverbial Buddhist Monk who was fond of angling and hence hooked a large fish and then left it into water as his Faith prohibited killing. Kaplan thus is suggesting that Russell's theory of descriptions is redundant and that Russell's effort is an exercise in futility! Nevertheless, it must be admitted that Russell does use expressions of the form (Ix)ϕx in positions reserved for terms in the non-definitional assertions of Chapter 14 of his (3). But in these cases (Ix)ϕx occurs not as a term of the language of (3), but as an abbreviation of a much longer expression of that language. That Russell uses other meaningful expressions which do not figure in the vocabulary of that language should be clear from say 14.204 which reads as:

$$E!(Ix)(\phi x).\equiv.(Eb).(Ix)\phi x = b.$$

E! is not a predicate of the language of (3); nor is existence for Russell, a property. Predication as such involves existence, hence 'existence' cannot be a non-redundant predicate. This is precisely why from E!x, as Russell would say, (Ex)(x = x) can be analytically inferred. On the other hand, contrary to Kaplan's reading, Russell seems to be using E! as well as (Ix)ϕx as abbreviations of expressions that figure in the language of (3). That ≡, and not

$=_{df}$ [22], is the principal connective of 14.204 should reveal this. And this abbreviation is warranted by 14.01 and 14.02 whose principal connective is $=_{df}$. But for abbreviational purposes such warrantedness need not be, thus should not be, invoked because, as Kaplan rightly points out, an abbreviation is a purely syntactical affair.

However, in so far as these abbreviations are intended not as time or space savers in the language of (3), they seem to be geared to highlight the fact that expressions like $E!(\mathrm{I}x)\phi x$ can altogether be dispensed with in that language as it is rich enough without them and includes expressions which are truth-functionally equivalent to them. Equivalence here is contextual in the sense that expressions from the language of the Principia are truth-functionally equivalent to predicative sentences in which definite descriptions occur as grammatical subjects. Despite this clarification, it will have to be admitted that Russell does indulge in what intrigued Kaplan. To pick an arbitrary example, in 14.15 of (3), Russell uses $(\mathrm{I}x)\phi x$ as if it were a term. His reason for this is that it is proper, then it obeys

> "the same rules with regard to identity as symbols which directly represent objects;"

for instance note the way in which the variable y behaves in

$$\{(\mathrm{I}x)\phi x = y\} \rightarrow \{\psi(\mathrm{I}x)\phi x \equiv \psi y\}.$$

But no substitution based on 14.01 or 14.02 can be made in this sentence, for such a substitution results in an ill-formed, hence meaningless, expression, say

$$[(Ex)\{(y)(\phi y \equiv x = y) = y\} . \rightarrow . (Ex)\{(y)(\phi y \equiv x = y) . \psi x\} = \psi y].$$

This is not even an expression for it flouts the rules of formation[23]. No such anomaly crops if the right side of 14.204 is substituted for the left side of it in all of its apparent occurrences in the canonical language of (3), for these are eliminable occurrences and are there only for convenience. So wherever (Ix)ϕx occurs in the assertions of Russell (3), (Eb){(Ix)ϕx = b} can be substituted for it without altering the truth-value of any of these assertions. Thus if 14.01 is a rule of transformation, 14.204 is a rule of substitution. But it cannot be made use of in those contexts in which expressions of the form (Ix)ϕx occur in isolation and not in sentential contexts as in E!(Ix)ϕx; this rule of contextual substitution does not have universal application. Its application is restricted to cases where the whole of an atomic sentence is replaced by another sentence without altering the truth-value of the sentence in which such a replacement is made. Thus it is a rule of sentential replacement. Giving a new meaning to an old term, we may say that such substitutions are Salva Veritate Substitutions, SV-substitutions for short. The SV-substitution rule is restricted to predicative atomic contexts. Sentences like (Ix)ϕx = y thus fall outside its range of application, as these are not predicative but are relational sentences[24], identity sentences to be specific.

In the case of sentences like (Ix)ϕx = y, substitution is never sentential or salva veritate; it is termial, and again in a new sense of the old term, Salva congrugate. We may call this kind of a substitution SC-substitution, for contrast. What the SC-substitution permits is, first a mutual substitution of expressions of the form (Ix)ϕx and expressions which are "symbols which directly represent objects," this is to say that constants and variables, and secondly substitution only when the substituend and the substituted for, both figure in some antecedently given predicative sentences. Otherwise, from x = x which is true, we may be arriving at (Ix)ϕx = (Ix)ϕx which may be false. This means that SC-substitutions presuppose SV-substitutions, which in turn presuppose the rule of transformation as in 14.01. Much ingenuity is not needed for putting all these matters in a systematic fashion, to systematize them i.e. But it requires some care to note that such a system of rules will not be a part of the rules of the canonical language of Russell (3). It only shows how the whole of language can so to say be telescoped into this canonical language, and how it thus provides a heuristic for legitimately believing that this canonical language has adequate expressive power.

Kaplan, and Scott whom he follows on several points, want to turn that system of rules into a part of the canonical language of Russell (3); they want to turn say 14.01 and 14.02 into sentences in the canonical language of the Principia, and for this purpose they collapse the distinction between $=_{df}$ and $\equiv$ which is crucial for Russell. Kaplan thinks on this line, but unfortunately imputing it to Russell. His distinction between Russell's "theory of descriptions" and Russell's "theorems of descriptions" clearly betrays this. Kaplan's contention is that the latter could be preserved while rejecting the former, for instance by accepting the Frege-Strawson or the Frege-Carnap theory. But what are Russell's theorems of descriptions? Unless the set of these theorems is the same for Russell as well as Kaplan, he cannot be said to be providing an alternative to Russell's theory of descriptions. Does that set include

(a) $(\mathtt{I}x)\phi x = (\mathtt{I}x)\phi x$

or

(b) $(\mathtt{I}x)\phi x = (\mathtt{I}x)\psi x \;.\!\equiv\!.\; \chi(\mathtt{I}x)\phi x \equiv \chi(\mathtt{I}x)\psi x$?

That the set of the theorems of descriptions does not include these is central to Russell's argument; and the reasons for their exclusion form the given set are given in, if they are not the same as, the theory of descriptions he propounded. It is true, as Kaplan says, that the theorems of descriptions are formulated in a certain language, and the theory of descriptions is about those theorems involving a semantic analysis of the language in which they were formulated. Russell would agree; but on the issue of the syntax of that language there can be no agreement between him and Kaplan, specially on the syntactical categories into which the vocabulary of that language can be classified, and also on the rules of formation which produce and identify each expression with its unique syntactical category. If Russell does not include (a) and (b) above in the set of theorems of descriptions, though he includes $x = x$ and $x = x \equiv (\phi x \equiv \phi y)$ in it, it is because they do not belong to any syntactical category of the vocabulary of the language of his (3), least of all to the category to which x and y belong. (a) above can be included in that set only if there is some y such that it is of the appropriate syntactical category turning $y = y$ into a member of that set, and it is the case that $(Ey)\{(Ix)\phi x = y\}$ i.e. $(Ix)\phi x = y$ - is true; otherwise (a) is not a member of that set. Moreover, from the point of view of Russell's theory which involves a semantic analysis, = is a relation between objects; thus $\underline{a} = \underline{a}$, where $\underline{a}$ is a variable ranging over the individual variables of the language of Russell (3) as well as definite descriptions, will belong to the theorems of descriptions only if each of its substituends has denotation. If this is not the case, it cannot be true. Kaplan says that

> "when we see somebody seriously asserting a sentence which is either trivially true or trivially false, the only plausible thing is to assume that he does not understand it in such a way that it is trivially true or trivially false."[25]

However, he does not give the benefit of this methodological principle of understanding to Russell on the position he takes concerning (a) and (b). I do not hope that Kaplan believes that Russell asserted what he did assert jokingly! Unless we grudgingly even meet the author half way, it would be difficult for us to understand him, as Frege implores us to recognize; of course, we can ignore him and begin afresh, but then we should not give the impression that we are solving his problem in a better way. By turning (a) and (b) above, or for that matter 14.01 and other definitions of Russell (3), into theorems of descriptions, Kaplan and Scott have not rejected the basic insights of Russell, namely that non-denoting terms do not abide laws of identity and that a restructuring of the received grammar is required to handle them, but are following those very insights in reconsidering them as denoting terms, and thereby pullulating, at least nurturing, non-natural entities. For instance, Kaplan thinks that by turning 14.01 into a theorem of the (extended) canonical language of Russell (3), it would be possible to

> "even provide for true atomic sentences about non-existing individuals,"[26]

and this is precisely what Scott accomplished, as Kaplan notes[27].

This style of arguing is fraught with dangers; it is philosophically dubious. Yet, it is respectable as it has a long tradition beginning with Socrates. He seems to be arguing in Scott-Kaplan style, in the Elenctic dialogues at least. It is one thing to say that I have a hazy vision of the Eiffel Tower, and another to say that I have a vision of a Hazy Eiffel Tower; the latter does not follow from the former. Worse is that it does not provide, in any meaningful sense, a solution to the problem of haziness in perception[28]. Treating the description operator as a term forming one - of the language of the Principia, i.e. -, taking 14.01 and its allies as theorems of descriptions, accepting the extensionality thesis that $x = y \equiv (\phi x \equiv \phi y)$ while considering it as an unrestricted rule of substitution, and subscribing to denotationist semantics, together, push one to covet such entities. That is what has happened to Scott and Kaplan. Kaplan, no doubt, has a rather technical and at the same time a tricky reason for treating the description operator as a term forming one, and 14.01 as a theorem of Russell (3). In essence it is that this operator cannot be eliminated using the rest of the devices available in the Principia. While rejecting Quine's contention that 14.01 is logically at par with say $P \vee Q .\equiv. \sim P \rightarrow Q_{df}$, which eliminates the connective $\vee$, and that this and 14.01 are equally contextual definitions on the count that in either of them the definitum does not coincide with the definiendum, Kaplan writes:

> "In the case of 'or' the definition could be thought of a really introducing an explicit definition ... As long as we introduce just the new sign plus variables for the argument expression, it can be transformed into an explicit definition of just the new sign, provided that the logic available ... is strong enough. If we had the lamda operator, for example, you can take the variables over the other side, and just define the connective. I don't think that anything like that can be done for Russell's iota without changing its grammar from a term-maker to a formula-maker, because of the variables on the left;"[29]

this is to say that the variables in $\chi(\text{I}x)\phi x$ include the variables that occur in χ which is not an argument for the description operator. Thus all that Kaplan wants to say here amounts to that the canonical language of (3) is not strong enough to eliminate the description operator by an explicit contextual definition, and hence, to enrich its expressive power, the description operator will have to be introduced into it via an axiom. But such an axiom will not be a definition, least of all one which permits elimination of the symbol in question. Kaplan is aware of this, for he writes:

> "We should carefully separate different kinds of definitions because they have different logical purposes which are of some importance. The so-called implicit definitions aren't definitions in any sense. We have a number of primitive constants; write down axioms on them, and people say, "well, you have implicitly defined them." You have, if the system has only one interpretation. If it is categorical, then there is a certain sense in which one might say that they have been implicitly defined. But in general one hasn't defined them in any sense - they have just been left open."[30]

This in fact is the position of Frege against Hilbert which we mentioned earlier, Frege has a point despite Hilbert or Hintikka. But that is not as important as that if 14.01 is turned into an axiom (or a theorem) - by replacing = by ≡ - the variables to the left of and the variables to the right of it cannot be allowed to range over the same domains if it is to have some semantic import[31]. Otherwise the original 14.01 which was treated as a rule of translation will turn into a rule of syntactical transformation; thus arises the need to restructure the semantic realm by invoking exclusive and jointly exhaustive alternatives to handle the problems that arise in representing the world in language. For there is no perfect resemblance between the world and language, one needs to be attuned to the other; thus either one has to restructure language, or else has to restructure what it is intended to represent. This choice between the syntactical and the ontological solution seems to be not at all guided by logic (alone, at least) for it largely depends upon the metaphysics to which one committed[32].

Scott's system to which Kaplan refers to clearly shows what we have been complaining about. In it, expressions of the form $(\mathtt{I}x)\phi x$ are terms which indeed are eliminable; and this is so as the following holds in it:

$$\chi(\mathtt{I}x)\phi x \equiv (\mathrm{E}y)[\{(x)(x = y) \equiv \phi x\} . \chi y] \vee [\sim(\mathrm{E}y)(x)\{(x = y \equiv \phi x) . \chi(\mathtt{I}x) \sim y = y\}]$$

Why first of all introduce terms and then eliminate them! The only reason for this seems to be to gain some legitimacy for using expressions of the form $(\mathtt{I}x)\phi x$ prior to their being decided proper or improper. If you provide a value for the improper ones among them - in fact, a single value for all of them - by instituting an outer domain, the very distinction between proper and improper definite descriptions breaks down at least in so far as we understand this distinction in terms of the proper ones having denotata and the improper ones failing on that count. This is so because we can say that all definite descriptions have denotata, some of them in the proper domain and some in the outer domain (or the improper domain!). In addition to this, Scott's improper domain is rather queer; it just has one member in it, and this lone member is denoted by countably infinite number of terms! Even if this member is just a make-believe one and is not full-blooded - the technical term "virtual entity" does not infuse anything to it - it still makes one wonder how on earth formal semantics with all its virtuosities has solved the philosophical problem concerning denotation! Worse is that these high-powered techniques have not even "skirted" the problem; it still stands naked! To notice this one may consider Scott's schemata of elimination given above. Suppose that $(\mathtt{I}x)\phi x$ is improper in the Russellian sense of the term, then it reduces to

$$\chi(\mathtt{I}x)\phi x \;.\!\equiv\!.\; \{(Ey)\{(x)(x = y) \equiv \phi x \,.\, \chi(\mathtt{I}x) \sim (y = y)\}\,.$$

The right side of this equivalence still has a term beginning with the description operator. (It is a term on Scott's reckoning, but it looks like a sentence to others - but leave it.) Thus Scott's schema is not one which is geared to elimination, but only of substitution; in essence it just says that any improper definite description can be substituted for any other improper definite description, whereas such a substitution is not possible with definite descriptions with their denotata in the proper domain. Yet, Scott claims that his schema

> "gives practically a wholesale way of eliminating the descriptive operator ... (T)o eliminate an occurrence of $(\mathtt{I}x)\phi x$ we need only examine the context in which the term is found; we do not have to make our elimination depend on any peculiarities of the formula ϕ within the scope of the operator."[33]

If the complexity of ϕ is not its "peculiarity," Russell's elimination too does not require us to take into account anything concerning ϕ which Scott does not want to consider. But we may leave that aside to note that, given its semantics, Scott's system is one in which $(\mathtt{I}x) \sim (x = x)$ will not be eliminable; thus it will have at least one ineliminable definite description in it. Of course, he can eliminate that too, but only by the semantics he provided for it, and by replacing $\chi(\mathtt{I}x) \sim (x = x)$ by $\chi . {\sim}\chi .$ thus by taking recourse to something worse than Russell's alternative.

Despite this disappointing result, from what we have noted about it till now, it emerges that Scott's system is philosophically interesting. After all, what really matters in philosophy is how far a point can be pushed, to find out whether it breaks, or where it breaks. Scott, while subscribing to Quine's criterion of ontological commitment[34] and to the Russellian reading of the quantifiers and their relation to predication, existence, identity and a whole lot of his conceptual battery, wanted to reject 14.01 to push the extensionalist thesis to the wilt. His aim was explicit in the companion to his (1)[35]; he declares there:

> "our point of view is purely extensional, so the meaning should depend just on "values" of terms ... and not how they are defined (or written syntactically.)"[36]

To let this point of view envelop expressions of the form $(\mathtt{I}x)\phi x$, Scott expands quantification theory to include these expressions as terms, and allows them to figure in all axioms and rules; assuming, like Russell, that all predicates and functions are fully defined, augmenting the domain of interpretation by an outer domain, he tried to show that any sentence ϕ is equivalent to a sentence $\phi*$ without definite descriptions such that ϕ is provable in the extended system if and only if $\phi*$ is provable in the original system, and finally to establish that

> "the axiom of descriptions (i.e. $(y)\{y = (\mathtt{I}x)\phi x .\equiv. (x)(\phi x \equiv x = y)\}$) is only a rewrite rule which perhaps needs to be reformulated"
>
> as
> $\{y = (\mathtt{I}x)\phi x\} \equiv \{(\mathrm{E}y)\phi y \,.\, (x)(\phi x \equiv x = y\,)\}$.

This means that his aim is to have a non-conservative extension of the original system - which implies that the theory of singular terms is independent of the quantification theory, as van Fraassen would put it. And we have shown how his program fails on the count of $(\mathtt{I}x) \sim (x = x)$; this is to say that either this expression remains an ineliminable "term" of the extended system which thus will only be a conservative extension, or else his outer domain will be inutile, if not an exercise in futility.

Now, suppose that Scott succeeded in achieving his goal; how it would have affected Russell's conceptual structure! Surely, it would have vitiated the distinction between the category of names and the category of definite descriptions. This distinction is fundamental to Russell; if it is collapsed, his theory too would collapse. The one logician who recognized this centrality if the distinction is Robinson[37]. We may recall how highlighting this distinction Russell remarked in his (3):

> "It might be suggested that "Scott is the author of Waverley" asserts that "Scott" and "the author of Waverley" are two names for the same object. But a little reflection will show this would be a mistake. For if that were the meaning of "Scott is the author of Waverley," what would be required for its truth would be that Scott should have been called the author of Waverley; if he had been so called, the proposition would be true, even if someone else had written Waverley; while if no one called him so, the proposition would be false, even if he had written Waverley. But in fact he was the author of Waverley at a time when no one called him so, and he would not have been the author if everyone called him so but someone else had written Waverley. Thus the proposition "Scott is the author of Waverley" is not a proposition about names, like "Napoléon is Bonaparte"; and this illustrates the sense in which "the author of Waverley" differs from a true proper name."[38]

Packed in this passage are several important points, two of which are:

(i) 'Scott is the author of Waverley' is a predicatively indicative sentence, and not an identity,

and

(ii) Scott, it is a bare brute fact, was called 'Scott', and not 'the author of Waverley'.

(i) and (ii) are closely related; thus a rejection of one leads to a rejection of the other. Robinson thinks that there is flaw in (ii), and that this flaw is the outcome of a

> "tacit assumption that a name of a person or object can always be chosen arbitrarily; to put it the other way round, that the interpretation of a name is, initially, a matter of choice."[39]

Though Robinson admits "the force of (Russell's) line of reasoning," he does not see any "compelling reason for" what he considers to be Russell's tacit assumption; of course, his suspicion was reinforced by the fact that

> "many writers, both before and since, have propounded theories which imply that a description indeed is a name. The fact that they could do so with impunity indicates that there is a flaw in Russell's argument."[40]

That the flaw evidently is in Robinson's argument can be shown even with the help of a *non sequitur* making us wonder how a logician could have extended that as an argument in a serious discussion. That the Americans could with impunity flout the sacred principle national sovereignty so many times during the last two centuries does not mean that the principle is dubious. That apart, it is doubtful whether Russell did hold that persons and objects are called by their respective names which are chosen arbitrarily; nor is it the case that he held that they have been given the names they have because they are (by nature) what they are, or because they were ordained or destined to be given those names. These issues, which were raked by Plato in Cratylus, are rather irrelevant from his point of view, because language is given and in it a name can be found associated with an object. Plato may have been called 'Plato' because he was broad-chested, and Aristotle may have been called 'Aristotle' because his parents found it musical. All these are facts of the world, and not of the language. The linguistic fact is that 'Plato' denotes Plato, or that Aristotle is called 'Aristotle'. In learning a language we learn this much and nothing more; as Quine would put it, by forgetting a name we forget a bit of language! Robinson, in any case, is not even concerned with the pre-linguistic issue as to how an object comes to be assigned a name; he believes that in the beginning there were names. Given this Genesis view, his problem was how to assign the objects given to us, and their shadows in the outer domain, to all the names that are available to us, if need be by coaxing or coercing names to share objects.

The motivation behind Robinson's effort seems to be not so much to provide a better alternative to Russell's theory as it is to differ from him on a crucial point that is at issue, and then work out an alternative which is at variance with all the alternatives provided by those who differed with Russell on that point. And Robinson does it with proclaimed cynicism about the whole affair; thus he confesses:

> "We may ask whether any particular solution of the problem of descriptions is the "best possible." However, I do not think that this question can be answered with finality. Indeed, given any particular feature of a formal language, we first have to make up our minds whether it is our aim (i) to make the language as effective a tool as possible for particular purposes such as the investigation of the physical world or the formalisation of a mathematical theory, or (ii) to give a precise explication of the corresponding feature found in natural language ... (A)s far as (i) is concerned, descriptions can be dispensed with altogether ... (A)s for (ii), a complete explication would have to take into account modal contexts and even the distinction between objective situation and subjective knowledge and it seems unlikely that all these can be covered by a single formalism."[41]

What then is the purpose for which Robinson has worked out his alternative? It is, though this is hidden behind a complicated model theoretic facade, to gain legitimacy to the assertion that there are meaningful undecidable sentences, this is to say that there are expressions which are meaningful but cannot be assigned truth-values initially and under a given interpretation. Thus it is perfectly legitimate to sue sentences which may contain as their terms non-denoting expressions; to put it differently, it is not necessary that terms be shown to be denoting prior to their figuring in sentences. This is the same as to say that semantics need not be given a priority over syntax, and that semantics of the kind of sentences in question be handled by reinterpreting the language, or by extending the given interpretation. Now let L be the usual first order language - this is to say that, a part of the canonical language of Russell (3) - augmented with suitable rubrics to turn expressions of the form $(Ix)\phi x$ as its additional terms, and let M be its model; let m be a many-one mapping of L onto M in the normal fashion. We can let m be the Russellian mapping of the unaugmented L onto M. No harm will come of this; in fact, it provides a contrast. Robinson assumes that the primitive base of L includes "enough individual constants," at least as many constants as one needs, to denote all individuals of M, and interprets quantifiers substitutionally. The assumption apparently is rather dubious, and entails a host of other equally dubious assumptions, for instance one about the size of the domain of M. It requires a commitment on whether infinity is to be interpreted generatively or as completed; in any case the assumption is far-fetched. The legitimacy of the "so-called" substitution approach - the qualification is Robinson's - hinges on the position one takes on these issues. Suppose one quantifies over the Reals, and accepts completed infinity while reading quantifiers referentially or objectually! Here there is already a departure from the Russellian framework, for Russell at one time - to be precise between his (2) and (3) - did think of substitutional interpretation and then abandoned it for reasons that need not be restated here as they are well-known and are well-documented in literature[42].

It is a commonplace that, given the standard model theory, a sentence S of L is satisfied by (or true in) M only if ~S is not satisfied by M, but is nonetheless meaningful for M. This means that the relation of satisfaction (or 'is true' as a predicate) is dependent not just on M and S but also on m mapping L onto M in virtue of inducing a correspondence between the terms of L and the individuals of M. "In other words," this is Robinson's discovery, "it defines denotation in M" of the terms of L. This is to say that the defining conditions of m determine or "constrain" a term of L having denotation in M [43]. According to Robinson, Russell and others missed what he discovered because in so far as individual constants (names) are concerned, the constraint is minimal and is not revealed by syntax. m maps the remaining terms of L, this is to say that expressions of the form (Ix)ϕx, onto M "uniquely, and without the intervention of the relation" of satisfaction, this is to say that indirectly through syntactical relationships[44]. Robinson thinks that it is uncertain whether the notion of identity involved in this unique mapping is a syntactical notion or else is semantic in its nature. So let us first attend to the way Robinson clarifies "the role of the identity."[45] According to him,

> "one correct definition of the identity ... is undoubtedly to conceive of it as the set of diagonal elements of M × M. ... '=' then denotes this relation and it is correct to say that ... M ⊨ s = t if 's' and 't' are terms which denote the same individual in M. But the identity may also be introduced by this condition so that M ⊨ s = t by definition, if 's' and 't' denote the same individual under correspondence m, which is again assumed implicitly, and this seems more appropriate in connection with the discussion of sentences which involve both descriptions and identity. ... (T)his definition has the unusual disturbing feature that it involves a proposition about symbols. This is masked in the notation "M ⊨ s = t," which seems to indicate that the formula ... 's = t', like other Wff, a statement about M. However ... once we have appreciated the situation, no harm will come of it. The purist who would write 'a' = 'b' will perceive the error by looking at the result. And if we interpret the 'is' in 'Scott is the author of Waverley' by identity sign in the sense of the second definition, then we obtain just what we want, i.e. that 'Scott' and 'the author of Waverley' denote the same individual."[46]

This way of handling identity is clearly a case of post hoc, propter hoc. To say that two members of the domain set are the same if terms denoting them respectively are identical is one thing, and to say that two terms are identical if their respective denotata are the same is a different thing. No harm may not have been seen to come of it, yet it is completely murky[47]. After all, we may be still smiling as the bad news has not reached us, and all the objections which Wittgenstein raised to this confusion may just be omens that do not forebode well. Whatever it might be, this much seems to be clear, namely that though Robinson is not talking about "the identity," he is really giving multiple notions of identity, if not actually offering a notion of relative identity. Worse is that he is not giving a definition of (the) identity, but only of fungibility. This comes out clearly from Robinson's argument. Denoting the same individual by terms 's' and 't' is not absolute; they denote the same under correspondence induced by m. Now suppose that m mapping L onto M results in (Ex)ϕx and (x)(y){ϕx.(ϕy$\rightarrow$x = y)}, with ϕ free for y, being true in M, then there is some constant a such that ϕa is true in M. Given this, we can extend (or is it, alter) m to result in m' which maps (Ix)ϕx onto M such that it corresponds to the value of x which is the image of a under m, and ϕ(Ix)ϕx is true in M under m'. In a routine fashion we can keep extending m' depending upon the complexity of the composition of (Ix)ϕx. That however is not the issue; but what counts is the issue concerning the semantic status of ϕ(Ix)ϕx, whether it is the case that a = (Ix)ϕx under m. There is no way to decide on these for this case[48]. This means that Robinson's identity theory is at variance with that of Russell; Robinson's notion of identity is also at variance with the notion of identity as it manifests in ordinary language. His identity, for instance, envelops the copula in say

Scott is the author of Waverley

which, for Russell, is a predicatively indicative sentence[49]. Robinson's point, to repeat, is that

> "(i)f we interpret the 'is' in 'Scott is the author of Waverley' by the identity sign in the sense of our second definition, then we obtain just what we want i.e. that 'Scott' and 'the author of Waverley' denote the individual."

But this is another case of the old Euxodan trick, namely assuming what needs to be established. 'Scott is the author of Waverley', in so far as the copula is the sign of identity, stands at par with 'The author of Waverley is Scott'; they are of the same logical form which is what Robinson wants. But neither of them is a sentence until after m is extended to result in m'; they gather the status of being sentences, this is to say that they acquire meaning, only with m'. Why should this be so? Robinson's answer is:

> " ... if a description is meaningful at all, its denotation is completely determined by the very sentences which establish that it is meaningful."

This is a confusing remark, to say the least. Robinson can talk about the meaning of a definite description's having or not having denotation; and he can talk about the meaning of a definite description, stipulating that 'meaning' and 'denotation' are synonymous. Or else he can talk of a sentence in which a definite description occurs as a term and its being meaningful, in the sense of its having a truth-value, and then say that the denotation of a definite description is determined by the sentences which fix its denotation, be they true or false. In this trivial sense the denotation of a definite description comes to be constrained maximally. And this is what Robinson seems to be saying, for according to him the sentences which bestow denotation on say $(\mathrm{I}x)\phi x$ are (i) $(\mathrm{E}x)\phi x$, and (ii) $(x)(y)\{\phi x.(\phi y \rightarrow x = y)\}$. Evidently, the semantic of these sentences is unproblematic, and is completely determined by m itself. Thus there exists a constant say a such that in the model in which these two sentences are true, ϕa is also true. Now for ϕ, 'the author of Waverley', a for 'Scott', 'Scott is the author of Waverley' is decidable under m. Thus there are two sentences which are orthographically identical but one of which is decidable even in m, and the other decidable only in its extension m'. But there is no criterion applying which we can tell for a given occurrence of 'Scott is the author of Waverley' whether it is a predicatively indicative sentence, or an identically indicative sentence. It may be recalled that Russell's basic differentiation between a name's figuring as a subject term in a sentence, and a definite descriptions' figuring in that way is geared precisely to secure a procedure for deciding this. This shows why Russell was right in his basic insight, namely that proper names and definite descriptions should be kept in separate baskets, and why the latter, be they proper or improper, should be analyzed even if no harm will come or no evil will befall in treating a definite description, when it is proper, as if it were a proper name[50]. Thus Robinson's reading of descriptions is only a variant of Russell's analysis; in fact it is the same but differing in the way it is stated. His determination of the denotation of a definite description is also contextual; only the predicate taking the definite description as its term happens to be metalinguistic. Thus he formulates it as a rule of deduction

$$\frac{(Ex)\phi x \qquad (x)(y)\{\phi x.(\phi y \rightarrow x = y)\}}{\lhd \{(Iz)\phi z\}}$$

and

$$\frac{\lhd \{(Iz)\phi z\}}{\vdash \{(Iz)\phi z\}}$$

where $\vdash$ is used in its usual sense, and $M \lhd t$ is read as 't has a denotation, or is interpreted in M' under correspondence $\underline{m}'$. Thus $\underline{m}'$ is a conservative extension of $\underline{m}$ because now there will be two kinds of theorems, one of which begins with $\lhd$, and the other begins with $\vdash$! But by conjoining these two rules - Robinson's proof procedure is normal and hence permits such a coupling - we may have

$$\frac{(Ex)\phi x \qquad (x)(y)\{\phi x.(\phi y \rightarrow x = y)\}}{\vdash \phi(Ix)\phi x}$$

What emerges from this is that if Russell's attempts amount to gaining transparency in the realm of extension for what is opaque in the realm of intension, Robinson's attempts amount to finding a way out from the anomalies in object language in the realm of metalanguage; and the discussion above must have shown how cost ineffective his attempt was.

Further, in so far as $\phi(Ix)\phi x$ is predicatively indicative, as it is for Robinson, there are formidable Russellian problems based on his theory of types which Robinson seems to be ignoring, and for which he has no satisfactory answers. Also, if it is predicatively indicative, it gets bogged in the Platonic problems of self-predication. When a type is defined as "the range of significance of a propositional function," and if it is accepted as a rule that whatever involves of a collection must not be one of the collection, as Russell does[51], it can be shown that taking $\phi(\mathtt{I}x)\phi x$ as a predicatively indicative sentence leads to typical ambiguity. Of course, it may be possible to skirt this ambiguity by holding that it is the first and only the first occurrence of ϕ in $\phi(\mathtt{I}x)\phi x$ which is a function term; but then it can be pointed out that this reading collapses the distinction between arguments of a function and its values[52]. And in so far as $\phi(\mathtt{I}x)\phi x$ is taken to be an identically indicative sentence, this is to say that if it is $(\mathtt{I}x)\phi x = (\mathtt{I}x)\phi x$, Russell will have no objection to Robinson's

$$\frac{(\mathrm{E}x)\phi x \qquad (x)(y)\{\phi x \,.\, (\phi y \rightarrow x = y)\}}{\vdash \phi(\mathtt{I}x)\phi x} \qquad \text{(i.e. } (\mathtt{I}x)\phi x = (\mathtt{I}x)\phi x\text{)}$$

But there is a rub, for Russell would not stop at that. He would expect that this schema of denotation be one which is validly inferable, thus it be the case that

$$\frac{\vdash \phi(\mathtt{I}x)\phi x \qquad \text{(i.e. } (\mathtt{I}x)\phi x = (\mathtt{I}x)\phi x\text{)}}{(\mathrm{E}x)\phi x \qquad (x)(y)\{\phi y \,.\, (\phi y \rightarrow x = y)\}}$$

This however is not permissible in Robinson's system because there may not be a decision procedure for settling whether $\lhd (\mathtt{I}x)\phi x$ in a given model $\mathtt{M}$; this is to say that whether $(\mathtt{I}x)\phi x$ holds in $\mathtt{M}$ is still in dark. Thus Robinson is in the camp of free logicians[53] who question the invertibility demand of Russell, and in the next chapter we will consider the alternative of the free logicians to Russell's theory.

If the foregoing has shown anything it is that it is not legitimate to hold that expressions of the form $(\mathtt{I}x)\phi x$ figure in the canonical language of Russell (3), or that there is something seriously imperfect with the way in which he defines them. Nor are the alternatives in which definite descriptions are introduced as a separate category of terms, eliminable or otherwise, fare any better than the Russellian language. The entire chapter 14 of the Principia is to be treated as an explanation about his canonical language, and not as a formal statement of a part of that language, or as a statement in that language. A confusion on this point is as old as Wittgenstein's (1), where he seems to have believed that the description operator is a primitive notion of the Russellian language, and that it was introduced contextually. At 5.451 and 5.452 of his (1) Wittgenstein remarks:

> "If a primitive idea has been introduced, it must have been introduced in all combinations in which it occurs," and
>
> "The introduction of any new device into the symbolism of logic is necessarily a momentous event. In logic a new device should not be introduced in brackets or in a footnote with what one might call a completely innocent air ... But if the introduction of a new device has proved necessary at a certain point, we must ask ourselves "At what points in the employment of this device now unavoidable?" and its place in logic must be made clear."

What Russell has been at a loss to make others see is that

(1) the description operator is not a primitive idea in the language of the Principia and its symbolism, (2) what one can do with it through a symbolism in which it figures as a primitive can be done in the language of the Principia, and having recognized this one can use it in open light, and (3) at no point in the Principia it is necessary to introduce it to conservatively extend it; and finally (4) the above three results make clear its place in logic.

Notes

[1] Kaplan (1), Scott (1) and (2), and Robinson (1).

[2] loc. cit., p.181; emphasis added.

[3] I trust that Scott is using this expression contrasting it with 'pragmatic meaning'. He claims that his alternative "combines the best features of Russell's and Frege's theories, and that it incorporates Russell's strategy of contextual definition and has all the virtues of simplicity."

[4] Rusza (1), Chapter 3.

[5] Note for instance how misleading is Kaplan's remark in his (1), p.278 that "(t)his ambiguity - that is the ambiguity concerning the primary or the secondary nature of the occurrence of a definite descriptive phrase - is regarded by Russell as a simple scope problem on par with the party invitation which reads "Bring your wife or come stag and have good time." And he introduces the terminology of primary and secondary (and by natural extension tertiary, quaternary, and so forth) occurrences to indicate the intended scope of the description in sentences of the given form. (Later in Principia Mathematica he introduces the more technically satisfactory device of scope indicators.)"

[6] Russell (3), Chapter 14, p.173.

[7] See Frege's explanation concerning sentences containing 'Odesseus' as their grammatical subject in his (3).

[8] Hintikka (3); see van Heijenoort (1) also.

[9] Hintikka (3).

[10] Hintikka thinks that on Frege-Russell view of the matter "as metalanguage is impossible so is metalogic." This may be too strong a point to attribute to Russell, at least. Russell was aware of the distinction between language and metalanguage; see his (5). But he did not make use of it while revising his (3). This may be due to the reason that he subscribed to the universality of logic; this is to say that it may be due to the reason that he believed that a language and its metalanguage do share the same logic.

[11] Russell (3), Vol.1, pp.11-12; emphasis added. The second sentence should throw much light on his view of "propositions," and also his aversion to truth-value gaps.

[12] Unless the indefinite article is allowed to figure explicitly the sentence does not sound natural; and when it is allowed to figure in that way, the sentence can be said to be predicative and not an identity.

[13] Chapter III, p.72.

[14] In view of the fact that every consistent set of sentences has a model this may as well be formulated to read that every consistent set S of sentences is a language.

[15] Hintikka, loc cit., p.4.

[16] See Frege (4).

[17] Note the indefinite article; not all non-Euclidean geometries can be mapped onto the Euclidean geometry.

[18] Russell (3), Vol.1, p.11; second edition.

[19] loc. cit., p.284; incidentally van Fraassen's supravaluational evaluation fails on this count.

[20] op. cit., pp.282-283.

[21] loc. cit., p.285.

[22] Russell is sensitive to the difference between ≡ and =; in fact, it is because of this he struggled hard to define = in terms of ≡. An indifference to their difference or a conscious blurring of the distinction between them is what Kaplan seems to have taken from Frege. Frege has four criteria for an expression to be a proper name. These are (i) it does not begin with an indefinite article, (ii) it does not contain free variables, (iii) it cannot properly occur as a predicate in a sentence, and (iv) it can occur on the left or the right side of the identity sign. Sentences are, for Frege, proper names on the count of (i) - (iv). Their being so on the count of (iv) depends upon blurring the distinction between = and ≡, identity and material equivalence, i.e.

[23] Something like this was on my mind when in my (1) I thought that Russell's theory violates the categorical distinction between sentences and terms. I now realize that I was wrong.

[24] Considering them as predicative sentences one may fall into the traps of Plato and Leibnitz, and secondly confine oneself to the Peano-limits of reasoning.

[25] op. cit., p.290-291.

[26] op. cit., p.286.

[27] Scott's virtual objects, Kit Fine's arbitrary objects, Parsons' and Routley's non-existent objects, Barwaise's situations ... are some such non-natural entities with which philosophers have recently enriched the world.

[28] What could be the logical and the ontological relationship between the Eiffel Tower and the Hazy Eiffel Tower! The relationship which the model theorists project between the world, or the domain on the one hand and the outer domain on the other hand can be likened to this. Even the accessibility relations of which the possible world semanticists talk about much are as legitimate as this relation.

[29] op. cit., p.292.

[30] ibid.; emphasis added.

[31] This is one of the objections which Wittgenstein raised against Russell's definition of identity; see Wittgenstein (1) and Part Two of this volume.

[32] It may be argued that this is guided by consciously or unconsciously held or rejected theological principles, because it all boils down to whether in the beginning there was the world or the word! Those with strong will and stronger muscles may join Goethe in holding that in the beginning there was neither of these, instead there was deed.

[33] op. cit., p.199

[34] See Rao (2).

[35] See Scott (2).

[36] Emphasis added.

[37] Robinson (1).

[38] Chapter III, p.67.

[39] op. cit., p.493.

[40] ibid., p.493.

[41] pp.493-494; emphasis added. As far as (ii) is concerned, despite mountains of modal logical literature, there is only despair, and not much is gained since 1905; some modal logicians however are honest enough to admit this. Huges and Cresswell for instance remark towards the close of the chapter on "Identity and Description in Modal Logic," in their (1):

> "In this chapter we have reached fewer firm conclusions than usual; but the topics of identity and description are among the most difficult in modal logic and in the present state of the subject are still full of obscurities and unsolved problems."

To a large extent this despair is the outcome of pushing Russell's theory and analysis beyond their intended explanatory goals, and this, as Hintikka aptly puts it in his comment on Kaplan (1), is "bad theory and bad philosophy."

[42] Grattan-Guinness (4), Consuegra (2), and Hylton (1).

[43] This model-theoretic profundity means no more than the pedestrian triviality that Scott, for instance, is denoted by 'Scott' because it (he) was called 'Scott', or the philosophical invention that 'Scott' refers to Scott in virtue of its Baptismal connection with him. s

[44] By Kaplan's criterion, Robinson's formal system is logically imperfect because it involves two-level semantics.

[45] Emphasis added.

[46] ibid., p.495.

[47] For further clarifications on this point see Part two of this volume.

[48] This is precisely what makes Geach argue that identity is relative. See Ch.7 of his (2), and also Griffin (1).

[49] The Russellian analysis of this sentence is accepted and uncompromisingly defended perhaps only by Geach; see his (1), sections 36 and 75.
[50] His use of proper definite descriptions here indicates that he treated the rubrics of elimination supererogatory, that is as something good to indulge in but not bad not to do that. It even shows his respect for the received language and the received linguistic practices.

[51] See Russell (10), p.25 and 63. Russell's type theory is semantic and ontological in its orientation, and for this reason became the subject of Wittgenstein's attack in his (1); see specially 3.1. to 3.333. Wittgenstein is in agreement with the essential point of Russell; only he wanted it to be shown forth by syntax itself. As he remarks in 3.333 of (1):

> "The reasons why a function cannot be its own argument is that the sign for a function already contains the prototype of its argument, and it cannot contain itself. That disposes of Russell's Paradox."

Wittgenstein thus wanted to push Russell's move from semantics to syntax to its logical end, and wanted to achieve what Russell gained by his ontological notion of types within the realm of syntax using the notion of perspicuous syntactical types (categories). Pace 3.1 of Wittgenstein (1), namely

> "It must be possible to establish logical syntax without mentioning the referent of a sign ..."

[52] This could be what Wittgenstein had on his mind when he wrote to Russell - see Appendix to Wittgenstein (3) - that the use of variables in 14.01 of Russell (3) is not in order. But we may not harp on that because, as Menger (1) argued in Chapter II "Wittgenstein on Formulas and Variables," Wittgenstein's views on the notion of a variable have not crystallized even by the time he came to compose what we now have as his (2) and (5).

[53] In fact in the last and the 13th section of his (2) he sketches a system of free logic.

Chapter Five : Free Description Theories

Since the sixties free logicians have been offering alternatives to Russell's theory; and as these are too many to be considered individually we may allow them to be represented by their Homer, if not Hesiod, namely Lambert[1]. Their alternatives are too divergent. As a consequence of the rather belated lowering of the discipline, there are numerous incompatible voices, each claiming to be one of the apostles of language, and its nature and logical structure, and all claiming for more space and acceptance. But all this goes on against the background of Russell and his theory[2]. In fact what unites all the free description theorists is their acceptance of a part of Russell's analysis of the import of definite descriptions, and their decision to differ with him in part, but each in his own way.

If Robinson thought that Russell's argument towards the view that proper names and definite descriptions are of different categories is based on a false assumption, and hence is not acceptable, Lambert thinks that Russell's argument is equivocal. However, in the case of Robinson, we know the argument of Russell to which he was referring; but in the case of Lambert, there is no such clarity. Thus Lambert's argument is confusing, and as I see, this is not due to any inadvertency on his part. It seems to be his strategy to show his opponent in pale light. Consider, for instance, the following lines:

> " ... (O)ne might think that dissatisfaction with Fregean and Russellian theories of singular inference served as the primary motivation for free logic. Yes and no. Dissatisfaction can be direct or indirect. Direct dissatisfaction arises from concern over whether a given theory is vindicated by its aims and validated by the evidence for it. For example,

> Russell's theory of definite descriptions ... is thought by many to be vindicated by its aims - for instance, the elimination of a need for non-existent objects - though not validated by its sources of support - for instance Russell's 'proof' in Principia Mathematica that definite descriptions are incomplete symbols. Indirect dissatisfaction arises typically from independent construction of a theory for a given end which, upon further consideration, turns out to be incompatible with an already existing theory ... Historically much of the dissatisfaction free logicians have expressed for theories of the Fregean and Russellian kinds have been indirect ... Of course, now and then some free logicians shared their disagreement with Russell - Leonard, Hintikka, and Schock notably - but to my knowledge no one has ever begun his development of free logic with a direct critical analysis of the traditional approaches of Frege and Russell."[3]

Let us ignore the genetic or the developmental talk - what Lambert says on these issues seem to be only partly right, as I have shown in my (6) - and consider only the substantive and conceptual issues which this passage raises. One of the central points which it seems to be highlighting is that Russell has not - in his Principia, i.e. - supported his contention that proper names and definite descriptions are of different categories. Russell did give in his (3) several arguments towards that end. Some of these are logical and other epistemological; some of these are syntactical, and others semantic. Unfortunately, Lambert has not, in any of his writings, considered these arguments to show that the strongest of these arguments is insufficient to warrant Russell's contention that proper names and definite descriptions are of different syntactical categories. Worse is that he has not considered any of the arguments given by Russell in his (3). What Lambert takes into account in order to show that Russell's argument commits the fallacy of equivocation is what Russell remarks in his (12). Of course, in this remark, Russell makes use of one of the arguments he gave in his (3), to be precise, one of those he gave in his introduction to the second edition, to which Robinson too has referred. What Lambert thus considers as Russell's "proof" of his thesis is the following remark:

> "The central point of the theory of descriptions was that a phrase may contribute to the meaning of a sentence without having any meaning at all in isolation. Of this, in the case of descriptions, there is a precise proof: if 'the author of Waverley' meant anything other than 'Scott', 'Scott is the author of Waverley' would be false, which it is not. If 'the author of Waverley' meant 'Scott', 'Scott is the author of Waverley' would be a tautology (i.e. logically true) which it is not. Therefore, 'the author of Waverley' means neither 'Scott' nor anything else - i.e. 'the author of Waverley' means nothing. Q.E.D."[4]

In any case Lambert's presumed discovery of equivocation in this passage is not all that original; in an earlier chapter we referred to Coddy's discovery of the same in his (1) five years earlier - and since then several philosophers have cited that passage towards a similar end. Neither the passage itself, nor Lambert's comments on it should bother us much; but it needs to be noted from it that Lambert is not willing to grant to Russell the laxities which he himself indulges in concerning the term 'meaning'. After all, he could understand without surprising anyone what Russell meant; he states that Russell

> " ... counts both fulfilled and unfulfilled definite descriptions meaningful, but considers ... neither kind of definite description - hence no definite description - a logical subject. This is the point of Russell's famous argument in Principia Mathematica that definite descriptions are not names."[5]

It is still not clear which argument of Russell (3) that Lambert is referring to. However, the strongest and the logical evidence that Russell rallies round to sustain his segregation of definite descriptions and proper names seems to be one which he spells out on p.68 of the second edition of (3); and it is that, given 14.01, it can be seen from the right side of it how

> "(Ix)(ϕx) has completely disappeared; thus (Ix)(ϕx) is merely symbolic, and does not directly represent an object, as single small Latin letters are assumed to do ... (and) ... although "(Ix)(ϕx)" has no

> meaning by itself, it may be substituted for y in any propositional function Fy, and we get a significant proposition though not a value of Fy."[6]

But this is not the case with a proper name, say a, More importantly, the "proof" which Russell is offering here is "inductive" in so far as it is arrived at matching the principle and evidence - the phenomenon which the principles are intended to capture being the language as it has been given to us. Naturally this is not unexpected of Russell whose firm commitment to realism and empiricism are well-known; his logical taxonomy is thus a part of his logical zoology. In a sense, it is this background commitment to realism and empiricism which endowed Russell's theory the fecundity which has come to be attached to it. Though Russell thought out this theory as a solution to some basic problems he encountered, and about which we have said enough, he soon realized its philosophical significance and paid considerable attention to it only to subsequently regret, as it lead to serious misunderstandings of his theory. He noted how the sort of the analysis it yielded in foundational studies can be extrapolated to our knowledge of the external world to show that the things which we encounter in our daily life are really "fictions," or constructions out of units of experience. Thus Lambert is only partly right when he thinks that

> " ... historically speaking, pressing philosophical considerations rather than an abstract regard for the correct patterns of reasoning involving definite descriptions more often than not, motivated various traditions in logical treatment of definite descriptions. Part of Russell's motivation, for instance, was the desire to "demolish" the need to admit beingless objects such as Mill's round square (popularized later by Meinong) and the golden mountain ... (And) ... free description theories are motivated principally by the attempt to accommodate basic intuitions about the relationship or the lack there between logic and ontology."[7]

Of course, Russell's ontological concerns are too evident; but what is too often either missed or ignored is that in his view of things the ontological status of logic itself has an overriding influence on what he did to or with those concerns; when no such status is granted to logic, and instead the logical is located in the epistemological realm, the relationship between logic (and thus language) and ontology becomes tenuous if not absent; thus language becomes a mechanism in which we talk without talking about! Realistic commitment and empiricist affiliations are evaded by most of the free logicians only to covet idealism and rationalism, often unknowingly[8]. In any case, the claim of Lambert on behalf of the free logicians that they are only trying to accommodate some "basic intuitions" about logic and language becomes suspect when it is noted that Russell too was working towards the same end, and it also reveals in full light how these disparate basic intuitions are only tips of icebergs. Beneath the surface are huge metaphysical mountains.

Let us now get to the details about the modus vivandi and the modus operandi of the free logicians. First, they blur the distinction between these; to put it differently, they accomplish their task by assuming their aim as an accomplished fact. Their gripe is that the Russellian frame of things, predication entails existence, and uniqueness presupposes existence. Despite their effort to "prove" how this is obtained in Russell's framework, it is no news to anyone familiar with Russell's writings. He himself explicitly says that these are his tenets, and explains at length as to how they are characterizations of our basic intuitions into the functioning of language in its representational role[9].

Lambert, in his A Theory of Descriptions, sees all this as all that is defective in Russell's theory, because - given the definition in 14.01 - $\chi(\mathrm{I}x)\phi x \rightarrow (\mathrm{E}x)\phi x$, thus $(\mathrm{E}x)\phi x$ itself, and $\chi(\mathrm{I}x)\phi x\ (\mathrm{E}y)\{(x)\phi x \rightarrow x = y\}$ can be proved; but that these can be proved in Russell (3) is known to everyone including Russell and Lambert. Thus, in one sense, Lambert's "demonstration" of that in the framework of Russell's theory in the following, namely

(i) " ... any sentence containing a description containing 'ϕ' logically implies the sentence that there are ϕ's," and

(ii) " ... any predicate is exemplified,"

on one hand, and

(iii) " ... any sentence containing 'ϕ' logically implies the sentence that there is at most one thing that is ϕ,"

on the other, can be proved does not throw any new light. These consequences are all too evident from several theorems that Russell spelled out in (3), for instance 14.21 which reads as: $\psi(\mathtt{Ix})\phi\mathtt{x} \rightarrow \{\mathtt{E!}(\mathtt{Ix})\phi\mathtt{x}\}$. But Lambert's demonstration is illuminating in another sense; and this sense is important philosophically. It exhibits a different way of projecting the relationship between, or exhibits the independence of, the basic concepts looming large in the mechanism of linguistic representation. Standard quantification theory stands on the nested conceptual triad of existence, identity and uniqueness; it is based on the assumption that all these three are required for a successful representation of the world in language. Thus, within the framework of Russell, uniqueness requires identity, and identity needs existence. What Lambert is suggesting in the constructive part of his argument is that this conceptual map is wrong, and that though uniqueness requires identity, identity does not need existence; nor is existence required for representation! Further, as far as Russell is concerned, not only existence and identity, but also individuation and identity on the one hand, and individuation and predication on the other - thus individuation and existence - go hand in hand. Lambert however is of the view that individuation and existence are independent of identity, and hence we can meaningfully talk of the identity or the difference of the non-individuated and the non-existent as much as we can meaningfully make predicative statements about the non-existent. It is these "deeper conceptual issues" - or if one does not mind, these deep seated metaphysical and ontological faiths - that are effecting the differences among the free logicians on the one hand, and the differences between them and Russell on the other. The disagreements over certain theorems or theses that need to be included or excluded from the formalized system of logic are only surface manifestations of the radical rifts beneath. Thus, if Lambert thought that the deductibility of $(\mathtt{E}x)\phi x$, or $\chi(\mathtt{I}x)\phi x \rightarrow (\mathtt{E}x)\{(x)(\phi x \rightarrow x = y)\}$ in Russell (3)[11]

> "demands a rejection of, or at least a change in, Russell's definition in Principia 14.01,"

it is because, having rejected the conceptual basis of that definition, he could not have accepted the consequences.

Lambert's or others' free description theories are not motivated by a desire to "entertain the unconditional acceptance of certain "natural" theorems that, in Principia, are assertable only under restriction," for example $(\mathtt{I}x)\phi x = (\mathtt{I}x)\phi x$. These are accepted by them as theorems and are found natural, for they are informed by their intuition that the underlying conceptual base of language is the way in which they see it, but is at variance with the way Russell sees it[12]. Free logic thus needs to be understood as a formalisation of the received language so understood. As such the differences between the free logicians and Russell are not just on the count of the differences in their respective aims and the uses to which logical analysis and construction are oriented. Whereas the free logicians and those that follow them are interested in " formal representation of language," Russell struggled to overcome the problems that in "representation (of the world) in formal language", in other words, a language whose structure is transparent and unambiguous.

Though the free logicians differ with Russell as to the relationships that hold between the concepts of existence, identity, and uniqueness, they subscribe to his views as far as the content of each of these concepts is concerned. They are thus in agreement with him as to the way in which each of these concepts can be formally captured within the quantification theory. As we shall return to this point soon, here we may just note that the domain of the existents is understood, in both the frameworks, as the domain of the quantifiers; and identity is understood in either of these frames in the (pseudo) Leibnitzian way[14]. This is to say that the identicals are taken to be having all their properties in common. Uniqueness is captured by Russell as well as the free logicians through quantifiers which receive existential reading; because of this the free logicians do not have any grouse against 14.02 of Russell's (3), namely

$$E!(Ix)\phi x . = :(Eb) : \phi x . \equiv_x . x = b \text{ Df} .$$

However, having collapsed the distinction between proper names and definite descriptions, or having accepted that the latter too are primitive singular terms, the free logicians replace = by ≡ in the above definition, and turn the right side of the equivalence into the necessary condition for the purpose of $(Ix)\phi x$. Thus only 14.01 comes to be the bone of contention between them and Russell. In the free logical correlate of 14.01 too one half is acceptable to the free logicians; they accept that

$$(Ey)\{(x)(\phi x \rightarrow x = y) . \chi y\} \rightarrow \chi(Ix)\phi x,$$

but reject

$$\chi(Ix)\phi x \rightarrow (Ey)\{(x)(\phi x \rightarrow x = y) . \chi y\}$$

on the ground that predication does not entail existence. If this half too is accepted

$$\chi(Ix)\phi x \rightarrow E!(Ix)\phi x$$

comes out as a thesis, and forces the falsity of sentences containing improper definite descriptions, including self-identities like $(Ix)\phi x = (Ix)\phi x$, which as the intuition of the free logicians informs them, are true whether the description in question is proper or improper. Russell accepts such sentences as true only when the definite description occurring in them is proper. As he says,

> " ... such symbols as "$(\iota x)\phi x$" obey the same rules with regard to identity as symbols which directly represent objects. To this, however, there is one partial exception, for instead of having
> $(\iota x)\phi x = (\iota x)\phi x$
> we only have
> 14.28 $\vdash E!(\iota x)\phi x . \equiv . (\iota x)\phi x = (\iota x)\phi x$."[16]

Because identity and individuation go together, and because identity requires existence, the non-existent is Proteus-like eluding laws of identity and standing in need of individuation. The non-existent stands in need of completion, at least. So to say it is an incomplete entity crying for an analysis as much as a definite description is an incomplete symbol demanding analysis. With reference to such objects - in fact, with reference to all objects which still have to acquire objecthood, individuality, and particularity - it is possible to entertain that they satisfy fleeting sets of predicates. There is nothing wrong with the free logicians' belief that the non-existent can be subjects of predication; only they do not realize that any set of predicates can be flung onto each of them! Thus in so far as the presumed indication of $(\iota x)\phi x$ cannot be said to be satisfying a fixed set of predicates - this is to say that unless it can be said with certainty that $(\iota x)\phi x$ indicates the same object in the first as well as the second occurrences in $(\iota x)\phi x = (\iota x)\phi x$ - it cannot be taken as an expression which abides by the laws of identity.

Lambert's acceptance of $(\iota x)\phi x = (\iota x)\phi x$ as a "logical truth" whether the definite description figuring in it is proper or improper on the count that it is "natural" reminds us of Wittgenstein's pithy remark that logic is not as simple as logicians make it appear. To grant the status of a logical truth to it without the Russellian restriction is based upon the assumption that identity is not objectual, that it is only nomial, in fact termial, because definite descriptions too are treated as terms. And in giving them that status, the free logicians accept the very same relationship between identity and individuation to which Russell subscribed, namely that only the fully individuated can be either identical or different from (another) fully individuated entity. When $(\iota x)\phi x$ is improper, Russell deprived it even self-identity because identity for him is objectual, and the object required here is not yet individuated. On the other hand, if identity is termial, no such segregation of improper definite descriptions is required as they too are well individuated as are proper definite descriptions - individuated in so far as they are symbols and only symbols. On the nomial (termial) reading of identity, what individuates either of these kinds of descriptions comes to be the set of rules of formation of the terms themselves. In so far as the central thrust of the free logicians' enterprise is towards a formal representation of the received language, no harm will come to it except that sticklers of formalities may demand that the terms in question be kept under single quotes and thereby create problems. But let the free logicians get away with their practice with the excuse that Robinson offered[17].

The effort of free logicians towards a formal representation of language without placing any restriction on the rules of standard identity theory could have gained some significance, and could have received greater attention, if only there were a way to collapse the distinction between formal representation of language and representation in formal language[18]. And rubrics towards this end could have been sought if only there existed an isomorphism between language and the world as they are given to us. In the absence of that, the representation of the world in language will have to have reductions - reductions of the kind that can be noticed in Russell's theory. One the one hand language falls short of the world, and on the other it is richer than the world. This is an insight of Russell and Wittgenstein which is of immense significance. Missing it, the free logicians, gearing their logical activity to formal representation of language, had to random-walk in those parts of the received language which out-stretch beyond the world. Their internal differences over the semantics of definite descriptions is a case in point.

A free description theory was initially propounded by Hintikka in 1959 with a single axiom[19], namely

$$y = (\mathrm{I}x)\phi x .\equiv. \{\phi y . (x)(\phi x \rightarrow x = y)\};$$

and Lambert showed[20] that it yields

$$\phi(\mathrm{I}x)\phi x,$$

as a theorem. This was unpalatable to Lambert, perhaps because his intuition did not inform him it to be natural. Of course, he would not have any qualms in accepting it as long as the first occurrence of ϕ is a short for '$(\mathrm{I}x)\phi x =$', this is to say that as long as it is a self-identity sentence. But he would consider it unwelcome if it were to be a predicative sentence. Self-identity is all right, but self-predication! It will lead to contradictions and paradoxes. Russell's experience, and what he learned from it, has been accepted by the free logicians, at least by Lambert. Thus the need to avoid self-predication, and covet self-identity. In that direction, Lambert weakened Hintikka's axiom to

$$(y)\,[y = (\mathrm{I}x)\phi x \equiv \{\phi y . (x)(\phi x \rightarrow x = y)\}]\,.$$

This weak version has come to be called Lambert's Law. It is evident that Lambert's Law and Hintikka's axiom are not identical within the framework of the free quantification theory. In that theory, the validity of $(x)\phi x \equiv \phi x$, which is a theorem of the standard quantification theory, is denied. As the free logicians redeem classical languages from existence, this is to say that from the requirement that terms have denotation, the rule of Universal Instantiation gets restricted in its application. Thus ϕx cannot come out as a consequence of $(x)\phi x$ unless and until certain conditions are met. From Hintikka's axiom and

Lambert's Law can be derived. This law is designed to capture the necessary and the sufficient conditions for capturing uniqueness which has been delinked from existence. It is tagged to a free logic suitably extended to include the standard identity theory which preserves unrestricted reflexivity, symmetry and transitivity of the identicals which obviously are all and only terms. The theory so arrived at is called minimal free description theory. This minimal theory accommodates the natural intuitions of the free logicians concerning self-identity of all definite descriptions and thus differs from Russell's theory. As it also meets that end with reference to identity sentences in which proper definite descriptions figure, it agrees with Russell's theory. But it fails to accommodate identity sentences in which improper definite descriptions occur, this is to say that in the case of sentences like $(\mathtt{I}x)\phi x = (\mathtt{I}y)\phi y$, where both the terms are improper, for instance 'The King of India in 1997 = The Queen of USA in 1979'.

This minimal theory stands in need of an extension if sentences like the one above are to be accommodated in it, which is the same as to say that if they were to be assigned truth-values; and it is in working out this extension that all the internal differences among the free logicians come to the fore. Augmenting the free quantification theory with identity and Lambert's Law is equivalent to assuming

$$\mathtt{E!}(\mathtt{I}x)\phi x \rightarrow [y = (\mathtt{I}x)\phi x \equiv \{\phi y . (x)(\phi x \rightarrow x = y)\}]$$

instead. This is to say that within the framework of free logic, Russell's analysis of proper definite descriptions as well as the way in which he suggested that they be eliminated hold good, at least in the sense that they seem to be articulations of some basic intuitions concerning language. However, there is no unique and well-defined basic intuition of free logicians concerning improper definite descriptions; thus alternative and often mutually incompatible extensions of the minimal theory have come to be developed. For example,

Lambert himself suggested two alternatives for extending the minimal theory to result in two non-equivalent theories. One of these referred to as FD1, is the result of adding to the minimal theory

$$\vdash (\mathrm{I}x)(x = b)$$

and the other, referred to as FD2, is the outcome of replacing Lambert's Law by

$$(\mathrm{I}x)\phi x \equiv (y)[t = y \equiv \{\phi y \,.\, (x)(\phi x \rightarrow x = y)\}].$$

When this schema is added to the extended free quantification theory, Lambert's Law can be derived as a theorem[21]. FD2 is equivalent to Scott's free logic, which we mentioned in the previous chapter, together with Lambert's Law. The rub with FD2 kind of theories is that they identify all improper definite descriptions, and thus assign the truth-value 'truth' to all identity sentences in which figure two different definite descriptions which are improper. These theories thus require that all improper descriptions be assigned the same referent. In this sense, they are Fregean in motivation and spirit. Of course, for mathematical purposes - for "reconstructing" it, as if mathematics is in a desperate need for reconstruction - no harm will come of this, for improper definite descriptions can be dispensed with altogether. Indeed for this purpose all definite descriptions can be dispensed with in terms of function terms in the familiar way. But for the purpose of representing the logic of ordinary language, FD2 may sound odd as it does not match our intuitions concerning the behavior of improper definite descriptions in day to day discourse.

Frege of course had a condescending attitude towards the ordinary language and consequently wanted to improve it by almost imposing on it the logical structure he invented. But Russell's attitude towards ordinary language is rather sympathetic. This is one of the reasons for which he rejected Fregean semantics and developed his own alternative of which his theory of descriptions is a part. As our concern here is Russell's theory and not free description theory as such, we shall not pay anymore attention to it, all the more so as we have come to notice that it is more an alternative to Frege's theory than to Russell's. We shall restrict our attention to Russellian free description theories to see how they fare when compared to Russell's theory. Lambert thought that a free description theory which is Russellian in motivation, FDR to wit, can be obtained by augmenting the minimal free description theory with

$$\chi(\mathtt{I}x)\phi x \equiv (\mathtt{E}x)\{y = (\mathtt{I}x)\phi x \,.\, \chi x\} \,.$$

FDR has a large number of followers, and it has been adopted in one version or the other by Schock (1), Scales (1), Kaplan (1), and Smiley (1). A free description theory is said to be Russellian in the Kaplanian sense of the term if it succeeds, as Lambert puts it, in

> "yielding the same truth-value verdicts for statements including definite descriptions inherent in Russell's famous definitions in Principia Mathematica."[22]

This means that a Russellian free description theory is one which

(1) retains all the Russellian theorems containing proper definite descriptions, and

(2) retains all the Russellian theorems containing improper definite descriptions when the background quantification theory is replaced by a free quantification theory, as it hardly needs to be mentioned. Lambert's replacement of the background quantification theory amounts to introducing into the primitive base of the canonical language of Russell (3) a monadic predicate, say Q, and placing a restriction on the rule of inference, specially Universal Instantiation, such that the augmented Russell (3) with the restricted rule of instantation is related to the original system in a well- defined way; a sentence ϕ is a theorem of the original system if and only if $\phi*$ is a theorem of the revised system, where ϕ and $\phi*$ are recursively correlated in the following way[23]:

(i) $P(t_i,\ldots,tn)* = Pt_i,\ldots,t_n$

(ii) $(E!t)* = Qt$

(iii) $(\sim\phi)* = \sim(\phi*)$

(iv) $(\phi \rightarrow \psi)* = \phi* \rightarrow \psi*$, and

(v) $\{(x)\phi x\}* = (x)(Qx \rightarrow \phi* x)$.

Q does not figure in the vocabulary of the canonical language of Russell (3), and is introduced into its symbolism as an auxiliary predicate to explain or to drive attention to some feature of the rest of the notation, and also the notion of representation. In fact, within the framework of the canonical language of Russell (3), it can be defined without residue, just like E!, in terms of quantifiers and identity say by contextually capturing it as $Qt \equiv (Ey)(t = y)$. But in the unextended quantification theory, in the absence of identity, Q is indefinable and hence ineliminable[24]. In virtue of this, it is legitimate to hold that E!, Q, and = are predicates of a specific kind; and this should naturally cast a doubt on the free logicians' attempt to delink existence and identity. Further, it may be interesting to note that Q can be read as 'denotes', and that it can be read so without getting into serious difficulties[25]. Because of this the free logicians' attempt to delink the concept of denotation and identity in order to allow self-identity to non-denoting terms becomes suspect. By noting how the sentence E!t, which apparently is in the material mode, and the sentence Qt, which will be in the formal mode if it is read as 't has a denotatum', we will have realized how identity collapses this distinction in virtue of capturing both of them. This means that, saying that an object a is existent, and saying that the name of the object 'a' denotes are saying the same thing in different ways in so far as the activity of representing the world in language is concerned; and saying either of these is equivalent to saying that the object is one of the values of a bound variable, or that it is not identical to any of the values of some bound variable. Thus, as subsequently netted the essence of Russell's insight in the pithy remark:

> "There are no ultimate philosophical problems concerning terms and their references, but only concerning variables and their values, and there is no philosophical problem concerning existence except in so far as existence is expressed by the quantifier …"[26]

Free logicians on the other hand hold diametrically opposite and incompatible views; according to them, logic is (or is it should be!) absolutely devoid of existence assumptions with respect to terms, be they singular or general, but nonetheless quantifiers (should) have existential force. Within the classical framework of Russell, because predicates are exemplified or instantiated in the domain which is given and well-defined, ϕx, $(Ex)\phi x$, and $E!x$ all hold together, or else none of them holds. This means that the existential reading of the quantifiers goes hand in hand with the assumption that general terms are exemplified in the domain. And as existential reading of the quantifiers and the use of non-denoting singular terms in predicational contexts cannot go together, Russell dropped non-denoting terms and assigned falsity to all sentences in which non-denoting singular terms figure as subject terms. ϕt thus is false either when ϕ is uninstantiated or when t is a non-denoting term. In the classical frame of Russell, as both general and singular terms are loaded with existence, it becomes possible for him to define one in terms of the other, of course with the help of identity and quantifiers. When for instance ϕ is uninstantiated this fact can be captured by $\sim(Ex)\phi x$, and when t is a non-denoting term, it can be expressed as $\sim(Ex)(x = t)$. This is what Lambert would categorize as a "negative semantic frame" in which identity entails existence.

In the positive semantics of Lambert, which assumes the independence of identity and existence, ϕx may be true even if ϕ is unisntantiated, and its truth may be warranted by intuition; and ϕ may even be '$x =$' whence ϕx is logically true. Thus logical truth becomes essentially a matter of language - syntax and symbolism, to be specific - and has nothing to do with the world;

this is a much debated and still debatable issue[27]. In so far as this assignment of truth-value is guided by intuition into the naturalness of linguistic usage it is suspect, and surprisingly is so even for Lambert. He holds that

> " ... naturalness is not always a good measure of truth."[28]

But then, one can wonder, when is naturalness a measure of truth! Unless this issue is settled to the satisfaction of everyone, Russell's semantics cannot be set aside as a misrepresentation of language and its normal functioning. In the absence of any attempt in that direction, it may have to be thought that Lambert and other free logicians are motivated by concerns other than naturalness in their thinking about alternatives to Russell's semantics. For Lambert emphasizes now on one point and then another, and as different free logicians play on different themes, it is rather difficult to pin-point the real reasons for their dissatisfaction with Russell. Nevertheless it seems certain that it is, as Lambert himself thoughtfully comments, occasioned by "clashing philosophies of language;"[29] as he says,

> "different concerns call for different theories of definite descriptions."[30]

In view of the uncertainty concerning those concerns, it may not be unfair to think that the advantages which Lambert counts towards a free description theory itself to be his primary concern; and the only advantage he counts is that it enables us to treat all definite descriptions as logical subjects - an idea about which Russell thought and wrote enough. Yet, adding a footnote to what he said, we may ask: what is the sense in which this is an advantage? Does this in any way augment the expressive power of our formal language? Or does it enable us to represent the world better in some logical sense of the term? To repeat what we remarked earlier, the necessary and sufficient conditions for representation are that each term must have a denotatum, that it must denote the same object throughout the discourse, and that it should denote uniquely.

These three are the requirements of representation as such; and the standard quantification theory incorporates all these requirements. As such the world can be represented entirely in terms of generality, and singular terms may be introduced as dispensable addenda; as Wittgenstein would say ϕa means $(Ex)(\phi x . x = a)$[31]. This is what van Fraassen is emphasizing, of late. He argued at several places that the theory of singular terms is independent of quantification theory, and that this is the case in an integral part of the belief of the classical logician. Nor does redeeming general terms from these background assumptions makes sense, for when the primitive expressions whose having denotata is undecided, the basis of quantification itself will remain indeterminate; that is the message of Quine's innumerable writings across all these years. And redeeming general terms from existence assumptions means the same as setting individual variables free from existence assumptions; and the ways in which quantification theory is reformulated to that effect, as van Fraassen remarked, "have been faute de mieux for lack of adequate theory." In any case an adequate theory can be thought of only when one is prepared to relate the problem to the deeper issues in epistemology and metaphysics as in Russell. In the absence of it what we have are conceptual confusions covered by technical triumphs!

The theory of singular terms is either independent of the extended quantification theory or else it is not. Russell's point is that it is not only independent but even inutile for it hardly adds anything to the expressive power not provided for by quantification theory. Of course, it will be of some convenience because sentences containing singular terms can be used as abbreviations for other longer and complicated sentences of quantification theory. If, on the other hand, the theory of singular terms is not independent, then it will have to be shown that augmenting quantification theory with singular terms increases its expressive power; this is to say that it will have to be established that say $(Ex)(\phi x . x = a)$ and ϕa are not equivalent. I am not suggesting that it is impossible to do this; all that I intend to suggest is that any possible argument to that effect will invariably push us out of the bounds of extensionality because these can be different only intensionally. Thus, it seems that there is an unsustained assumption to which free logicians have not paid any attention before deciding to consider definite descriptions as possible singular terms. Of course, they can, as some of them did, say that definite descriptions are indispensable or eliminable singular primitive terms, but that would not be a serious joke either.

If the theory of singular terms is treated as independent of quantification theory, standard quantification theory turns out to be a sub-theory of free quantification theory, and if it is treated as a part of quantification theory, free quantification theory turns out to be a sub-theory of the standard quantification theory[32]; they are, as Lambert puts it, "now in abundant supply." But none of these is satisfactory as each of them incorporates some ad hoc device or an anomaly preventor. They either treat E! as an irreducible primitive predicate, or resort to vacuous quantification, or multiply domains of interpretation, or take recourse to some non-standard identity theory, or fall back upon some such strategy. If these technical tricks are evaded, and free logic is formulated as a sub-theory of the standard quantification theory such that the resultant system with an addition of a single axiom to the effect that $(Ex)\phi x$ coincides with the standard quantification theory, the resultant system cannot be extended to cover identity. This is to say that the standard identity axioms cannot be tagged on to it to issue in the extended quantification theory[33]. Identity axioms can be added to such a free quantification theory only when $(Ex)\phi x$ too is taken as an axiom, indicating that identity requires existence. And that clinches one the central theses of the free logicians and vindicates Russell. I said 'clinches' and not 'finally clinches' for in philosophy there are no final solutions or final rebuttals. Those who are opposed to a position or a standpoint will always be out to find the weak points in the outer walls to demolish the whole structure, and those who are determined to hold that position do their best to protect the strong corner stones to keep the architectonic in order against all attacks. The strength of a philosophical argument is to be measured, then, by the number of the attacks it withstands, and the variety in the directions of those attacks. On that count Russell's theory of definite descriptions comes out as a solid construction and still remains a paradigm of philosophical analysis.

Coda:

Lichtenberg once likened the philosopher to a dog to whose tail is tied a bone. He can smell it, but cannot see it; he can feel it, but cannot snatch it. So gets restless, almost neurotic. He starts moving in circles; and either due to his aimless activity, or as happenstance, the string becomes loose, and the bone falls right in front of him. That brings everything to rest. Russell was distressed by sentences of the form 'the such-and-such are such-and-such'. He smelled that they must be of the form $\psi(\mathrm{I}x)\phi x$. This in turn irritated him a lot. Then he had a flash, and realized that

$$\psi(\mathrm{I}x)\phi x = (\mathrm{E}x)\{\psi x \,.\, (y)(\phi y \equiv x = y)\}.$$

Do not, he warned himself, replace = by ≡ , for the former is richer than the latter; he pleaded that is has a bit of intensionality injected into it, whereas the latter is purely extensional. And understand = in terms of having predicates in common, he implored. Then you will see that the two sides of = here have all the significant semantic properties in common. He tried to convince that, in any case, these properties are only two in number - namely, meaning and truth. For all that we know, this is all that is there to it, as he believed.

Notes

[1] Lambert's writings are too numerous, and often repetitive; and the literature on free logic voluminous. See the bibliography in Rao (6). However, for the purpose at hand, acquaintance with Lambert (3) and (6) should be sufficient. The latter covers most of the important contributions towards free description theories, and contains, in Ch.2, his statement of his final position on the issue. For a brief survey of the much discussed free description theories, see Bencivenga (1).

[2] The same is the case with modal description theories; they are worked out under the shadow of Russell. See for instance Huges and Cresswell (1), Sylvan (1), and Slater (1). The one uncompromising Russellian modal definite description theory is in Smullyan (1).

[3] Lambert (3).

[4] Russell (12), p.84; it may be noted that he is not saying that there is, or that he has, a proof of his "thesis." From this it may be inferred that he did not intend this passage as a serious logical argument, but only as a heuristic to drive home a logical point.

[5] Lambert (3), p.58; emphasis added.

[6] Emphasis added.

[7] Lambert (3), p.57

[8] For a different argument towards the same end see Bencivenga (2) where it was shown how if classical logic is committed to transcendentalism, free logic is committed to idealism too.

[9] Almost a quarter century ago, in his (2), Hintikka suggested that what is required in this context is "deeper conceptual analysis of the situation than the standard syntactical and semantic methods afford;" but it seems that his suggestion is yet to be taken into cognizance by the free logicians.

[10] Ch.2 of his (6).

[11] There is a tricky issue in Lambert's way of deducing (Ex)ϕx in Russell's system; "letting χ be (λy){y = (Ix)ϕx}," he writes, "with the axiom of self-identity one can obtain (Ex)ϕx." Here the lamda operator is not Church's singular term producing one, but one which produces general terms. Thus it is more like Leonard's dot operator in his (2). (λx)Cx here is to be read as: the property characterizing an individual x if and only if x meets the descriptive condition C. Thus Lambert can think of substituting (λy){y = (Ix)ϕx} for χ in say χ(Ix)ϕx. But Russell's remark while concluding Ch.14 of his (3), namely

> "it should be observed that the proposition in which $(\mathtt{I}x)\phi x$ has the larger scope always implies the corresponding one in which it has the smaller scope, but the conserve implication holds only if either (a) we have $\mathtt{E!}(\mathtt{I}x)\phi x$ or (b) the proposition in which $(\mathtt{I}x)\phi x$ has the smaller scope implies $\mathtt{E!}(\mathtt{I}x)\phi x$,"

tantamounts to that the description operator always takes the smallest scope, which in turn amounts to that χ in $\chi(\mathtt{I}x)\phi x$ is atomic. Now replacing χ here by $(\lambda y)\{y = (\mathtt{I}x)\phi x\}$ brings in several confusions and anomalies; to mention only the obvious one, there is no way to indicate in the resultant sentence the scope of definite descriptions following the rules indicated by Russell towards that end. Thus $(\lambda y)\{y = (\mathtt{I}x)\phi x\ (\mathtt{I}x)\phi x\}$ remains syntactically anomalous, if not patently ill-formed; but we shall not harp on this point further.

[12] One can dismiss this olfactory explanation and justification by extrapolating arguments concerning intuitionist moral philosophy.

[13] The free logicians and Russell have different "basic intuitions" as to what constitutes the world. Russell with his "robust sense of reality" intuits the independent existence of the world consisting of, to put it platitudinously and following Quine, all and only what is there. This is to say that Russell's world has something positive about it; it does not contain negative elements, specially negative facts. Negation enters into the picture only in representing the world in language. The world makes up all and only grounds for true sentences being true, and thereby for false sentences being false. The world of free logicians, on the other hand, is a mere projection forming the ground for all those sentences which are informed by their intuition to be true; in so far as this intuition is linguistic, truth becomes a syntactical notion, and the world turns out to be reflection of a linguistic description of accredited veracity!

[14] See Part Two of this volume, and Rao (8).

[15] In fact Wittgenstein thought that Russell himself was confusing = and ≡ in order to sustain his logicist reduction of mathematics to logic; in his (1) at 6.22 he points out that mathematics contains equalities and logic consists of tautologies. Equalities and tautologies are not the same. If the differences between them is accepted, the logicist goal might be adversely affected; in any case it was shown by Godel to be unrealizable. Nevertheless the theory of descriptions can still be held to be the correct characterization of the semantic structure of our language, and this in no way is affected by the impossibility of realizing the logicist goal.

[16] Russell (3), Ch.14, p.175.

[17] See the previous chapter.

[18] In one sense Russell as well as the free logicians both are of the view that the standard quantification theory is in essence the logic of our language. Both of them hold that the angularities and the anomalies that arose and that may arise be smoothened out with some local restrictions on the rules of inference. Russell wants these restrictions to be placed on the rule of substitution in identity theory, and the free logicians want them to be placed on the rules of quantification theory, this is to say that on the rules of instantiation.

[19] See his (1). Of course, Leonard earlier suggested one in his (1), but it involves modal notions.

[20] See his (1); for Hintikka's reply to Lambert see his (2).

[21] Here we are assuming that while adding this schema as an axiom, the scope problems have been taken care of.

[22] See Lambert (3).

[23] See Meyer and Lambert (1).

[24] See Meyer, Bencivenga and Lambert (1).

[25] See the last section of Robinson (1), where he sketches his version of free quantification theory which is based on such a reading.

[26] Quine (1), p.224; and Rao (2).

[27] A similar idea was a part of the logicist belief shared by Frege and Russell. But Frege abandoned this life-long belief towards the close of his life. He came to think that logic alone cannot provide objects for which properties such as equality or set-membership can be appraised. See Frege (4).

[28] See "A Theory of Definite Descriptions," in his (6), p.20.

[29] Lambert (3), p.64.

[30] ibid., p.63.

[31] Part Two here, Ch.3.

[32] See Rao (6).

[33] See Rao (4).

Appendix : Rodriguez - Consuegra on Russell on Definition

Consuegra in his (3) traced in detail, with copious citations from Russell's published as well as unpublished writings, how (1) Moore showed Russell what it is to do philosophy, namely that it is the same as offering definitions in terms of simples which are indefinable, and how (2) Peano showed him how to do this using logical techniques. Though Consuegra, here as well as in his other writings, has thrown much light on several important points concerning the formative period of Russell's philosophical life, what he has to say about Russell's search for a plausible notion of definition needs some modifications. Apart from the minutiae that he has dug up from the unpublished works of Russell and the writings of Peano which are not as yet available in English (some of which are interesting to read in Consuegra's translation into English), he has, using those passages as his evidence, come to the conclusion that first Peano's theory of definition coincides "at least in essentials with Moore's," and secondly, though Russell was not happy either with Moore's or with Peano's notion of definition, he himself could not, in the Principles at least, offer a "convincing and complete theory of definition." Consuegra himself cited several passages from Peano's collected works which induce us to be skeptical about his notion of definition, even "in essence" being the same as Moore's. However we need not concern ourselves about it here. But as what Consuegra holds concerning Russell's inability to produce a theory of definition has bearing on my discussion here, I may attend to stating in brief what I hold on the point at issue. First, I do not know what sort of criteria that Consuegra expects a theory of definition to satisfy in order to be adjudged as a convincing and complete one. Secondly, I do not know why we should not consider Russell's characterization of a logico-mathematical definition in the Principles as one which embodies all the conditions which he thought to be necessary and sufficient for its being adequate. Of course, this characterization does not envelop all sorts of equalities which are passed as definitions either by philosophers or by the laity. I do not think that Russell ever thought of a characterization which would cover all of them and thereby have a claim for completeness in the sense of comprehensiveness, for then it is bound to be unconvincing, specially so as philosophical and logico-mathematical definitions are not of the same genus; nor do I think that Consuegra has on his mind such a global characterization when he talks about a complete theory of definition and its eluding Russell. Otherwise he would not have been unhappy with what he believes that he noticed in Russell, namely a confusion between philosophical and logico-mathematical definitions. In fact, his finding fault with Russell on that count should lead us to believe that even according to him, as it is according to Russell, there cannot be a common theory of logico-mathematical and philosophical definitions. Consuegra thinks that Russell is not clear (and perhaps, deliberately unclear) on whether (i) the definiendum is only an abbreviation of the definiens and whether (ii) the definiens is incremental in the sense that it analyzes the familiar but clouded idea which is present in the definiendum, and thus is "a notable advance." According to Consuegra, Russell is mixing up definitions as analyses and definitions as abbreviations, because he does not admit the distinction between nominal(ist) definitions of Peano brand and real(ist) definitions of the Lockean and Aristotelian variety. Consuegra is of the view that this is the case not with Russell of the Principles alone, because Russell of the Principia perpetuates the same vices. In fact, Consuegra holds that the Principia "is built on this ambiguity," and thus it is the result of Russell's attempt for "taking advantage of the logical techniques without getting stuck with their 'tautological disadvantages'." And, as Consuegra rightly points out, Russell could complete this attempt as he believed that his "logic can serve for technical definitions as well as philosophical ones;" here "logistic" would have been better than "logic", but it should not matter much, as it is well-known that Russell, either informed by Bradeley or Frege, set himself to discover this logistic by analyzing ordinary language in which philosophers normally offer their definitions. He believed that logical forms are hidden behind the most often misleading grammatical forms of ordinary language, and when extracted and exhibited are sufficient to formulate anything indicative, thus all known mathematical knowledge, and formulate it turning the real indication perspicuous. In any case, Russell legitimately believed that they are rich enough to express logico-mathematical as well as philosophical definitions. This is all commonplace to any student of contemporary philosophy. What then is the rub? It is, as I see it, in Consuegra's not recognizing that in the Principia, Russell is extending definitions in every possible (and accepted) sense of the term. Thus we have a logico-mathematical definition in 14.01 of the Principia, a philosopher's definition in 14.02, and nominal, constructive or stipulative definitions like $(P \rightarrow Q) \equiv_{df} (\sim P \vee Q)$ and $(P \equiv Q) \equiv_{df} (P \rightarrow Q) . (Q \rightarrow P)$. Russell did not, nor should we, think that all of them can be kept in the same basket, be it logical or epistemological. The third kind of definitions here are Peano-constructive, for according to Peano a constructive definition must have the same variables both in the same definiendum and the definiens. And they are nominal definitions too, as the definienda are abbreviations and are "superfluous". There is, however, another sense in which definitions can be Peano-constructive; in this sense, a definition is constructive if it is of the form $X = A$, where A possesses "known-sense" and X "still bereft" of sense, as Consuegra would say, but in a more meaningful terminology, where A possesses a given well-defined sense, and X lacking in this sense. In this sense of the term 14.02 is constructive, because $(Ex)\ldots$ has a well-defined and precise sense, and $E!$ as a predicate has not been understood this precisely till now. But given this definition it can be replaced. 14.01 on the other hand does not have the same variables in the definiendum and the definiens; the variables figuring in the definiendum and the definiens do not have the same set of substituends. In fact the variables occurring in the definiendum, say x in $(Ix)\phi x$, does not have substituends in the normal sense of the term; and in this sense it is not a variable either. Further, which variables figure in the definiendum (apparently it may not show any) can be ascertained only after a long process of analysis which is to be carried out bracketing the variables that apparently figure in the definiendum. Russell sketched the process of this analysis, and showed how the variables that explicitly figure in the definiens are the variables that are implicitly present in the definitum. He did this in the Principia, but he noticed that this indeed is the case in the Principles itself. In Sec.19 he wrote:

> "The mathematical definition does not point out the terms in question ... Only what may be called philosophical insight reveals it."

That the definitum of 14.01 is about what definiens is about, or that their indication is the same, is the outcome of a philosophical insight. Why then, Consuegra seems to be puzzled, Russell did say in Sec.31 that mathematical definitions do not involve analysis. The obvious answer is that Russell is right because insights are not analysis. But to be fair to Consuegra, we may say that mathematical definitions do not involve anything which has some affinity with Moorean type of analysis. The kind of analysis that they involve is not of complex concepts in terms of indefinable simples, but of logical forms - syntactical categories that go into the composition of the definitum. In so far as this analysis is geared towards making the syntactical form transparent, its importance is not epistemic but logical. The gains from philosophical insights to which Russell is referring are not in the realm of "meaning" but of grammar; they are not of conceptual understanding, but of syntactical formulation. To purge logico-mathematical definitions from these insights is to render them irrelevant. Unless Consuegra wants to turn the traditional High Table Toast (To Higher Mathematics, May it not be relevant to anyone!) into a serious philosophy of mathematics, these insights are of immense significance. In any case, what is meant by "philosophy" here is much wider and deeper than philosophy in the Moorean sense - namely, the activity of offering definitions of a sort. It may have some affinity with the basic design of Bradeley to penetrate through the grammatical structure to reach the logical beneath, but certainly is not oriented towards offering definitions. Of course, it goes towards legitimizing or justifying definitions which have been offered; its main purpose is to get syntax straight. Thus it makes no sense to talk about Russell's theory of definition being incomplete. And whether Russell's theory is convincing or not is to be seen in the global context of his endeavour of which the Principia sketches only a part. Thus it would be more appropriate to talk of his definition being adequate or otherwise. But then its adequacy is derivative of the adequacy of the language of the Principia; and this is a matter which is well settled.

<u>Bibliography</u>

Barwaise, J., (ed)
1. *Handbook of Mathematical Logic*, North-Holland Publishing Co., Amsterdam, 1981

Bencivenga, E.,
1. Free Logics, (in) *Handbook of Philosophical Logic*, (eds) Gabbay, D., and Gunthner, F., Vol. III, D. Reidel Publishing Co., Dordrecht, 1986.
2. Free from what? *Erkennitnis*, Vol. 33, 1990.

Beth, E.,
1. *Foundations of Mathematics*, North-Holland Publishing Co., Amsterdam, 1959.

Boer, S.E.,
1. Attributive Names, *Notre Dame Journal of Formal Logic*, Vol. 18, 1978.

Coddy, W.S.,
1. Russell on the Meaning of Descriptions, *Notre Dame Journal of Formal Logic*, Vol. 16, 1976.

Consuegra, F.A.H.,
1. The Origins of the Theory of Descriptions According to the Unpublished Manuscripts, *Russell*, Winter 1989-1990.
2. Russell's Theory of Types 1901-1910: its Complex Origins in Unpublished Manuscripts, *History and Philosophy of Logic*, Vol. 10, 1989.
3. *The Mathematical Philosophy of Bertrand Russell - Origins and Development*, Birkhauser Verlag, Zurich, 1991.

Donnellan, K.,
1. *Reference and Definite Descriptions*, The Philosophical Review, 1965.

Dummett, M.,
1. *Realism and Truth*, Princeton University Press, Princeton, 1984.

Van Fraassen, B.C.,

1. Singular Terms, Truth-value Gaps, and Free Logic, *The Journal of Philosophy*, Vol. 67, 1966.
2. The Completeness of Free Logic, *Zeitschrift fur Logik und Grundlangen der Mathematik*, Vol. 13, 1966.

Frege, G.,

1. *Conceptual Notation and Related Articles* (1879), translated and edited by Bynim, Y.W., Clarendon Press, Oxford, 1972.
2. *The Foundations of Arithmetic: A Logico-Mathematical Enquiry into the Concept of Number* (1884), translated by Austin, J.L., Blackwell, Oxford, 1950.
3. On Sense and Nominatum, (in) *The Philosophical Writings of Gottleib Frege*, translated by Geach, P.T., and Black, M., second edition, Blackwell, Oxford, 1960.
4. *Philosophical and Mathematical Correspondence*, translated by Kaal, H., abridged by MaCguiness., and edited by Gabriel, G., Herman, H., and Kambartel, F., Blackwell, 1980.
5. *Posthumous Writings*, edited by Hermes, H., Kambartel, F., Kaulbach, F., and translated by Long, P., and White, R., Blackwell, 1979.
6. *Basic Laws of Arithmetic: Exposition of the System*, edited and translated by Furth, M., University of California Press, 1964.

Geach, P.T.,

1. *Reference and Generality*, Cornell University Press, Ithaca, 1968.
2. *Logic Matters*, Blackwell, Oxford, 1972.

Goldfarb, W.D.,

1. Logic in the Twenties: The Nature of Quantifiers, *The Journal of Symbolic Logic*, Vol. 44, 1979.

Grattan-Guinness, I.,

1. The Russell Archives: Some New Light on Russell's Logicism, *Annals of Science*, Vol. 31, 1974.
2. Russell's Logicism and its Influence, 1910-1930, *Proceedings of the Third International Wittgenstein Symposium*, Holder-Pichler-Tempsky, Vienna, 1979.
3. On the Development of Logic Between the Two World Wars, *American Mathematical Monthly*, Vol. 88, 1981.

4. From Principia Mathematica to Godel's Incompleteness Theorem, *History and Philosophy of Logic*, Vol. 5, 1984.

van Heijenoort, J.,
1. Logic as Calculus and Logic as Language, Synthese, Vol. 17, 1967.
2. *From Frege to Godel: A Source Book in Mathematical Logic*, Harvard University Press, Cambridge, Mass., 1961.

Hilbert, D., and Ackerman, W.,
1. *Principles of Mathematical Logic* (1928), English translation, Scribners, New York, 1950.

Hintikka, J.,
1. Towards a Theory of Definite Description, *Analysis*, Vol. 19, 1959.
2. Definite Descriptions and Self-Identity, *Philosophical Studies*, Vol. 15, 1964.
3. Existential Presuppositions and Uniqueness Presuppositions, (in) *Philosophical Problems in Logic*, (ed) Lambert, K., D.Reidel Publishing Co., Dordrecht, 1970.
4. Model Theoretic Viewpoint in Logical Theory, *Synthese*, Vol. 77, 1968.

Hughes, G.H., and Cresswell, M.J.,
1. *Introduction to Modal Logic*, Methuen, London, 1968

Hylton, P.,
1. Russell's Substitution Theory, *Synthese*, Vol. 45, 1980.s

Kaplan, D.,
1. What is Russell's Theory of Descriptions?, (in) *Physics, Logic, and History*, (eds) Yourgrau, W., and Beeck, A.D., Plenum Press, New York, 1970.

Kripke, S.,
1. *Naming and Necessity*, Blackwell, Oxford, 1983.
2. Is There a Problem About Substitutional Quantification? (in) *Truth and Meaning*, (eds) Evans, G., and Mcdowell, J., Oxford University Press, Oxford, 1976.

Lambert, K.,

1. Notes on E! - III: A Theory of Descriptions, *Philosophical Studies*, Vol.13, 1962.s
2. Notes on E! - IV: A Reduction in Free quantification with Identity and Definite Descriptions, Philosophical Studies, Vol. 15, 1964.
3. On the Philosophical Foundations of Free Logic, *Inquiry*, Vol. 32, 1981.
4. *Meinong and The Principle of Independence*, Cambridge University Press, 1983.
5. On the Philosophical Foundations of Free Description Theory, *History and Philosophy of Logic*, Vol. 8, 1987.
6. *Philosophical Applications of Free Logic*, Oxford University Press, New York, 1991.
7. A Theory About Logical Theories of "expressions of the Form 'The so and so' Where 'The' is Singular", *Erkennitnis*, 1991.
8. Russell's Version of the Theory of Definite Descriptions, *Philosophical Studies*, 1992.

Landini, L.,

1. Russell's Substitution Theory of Classes and Relations, *History and Philosophy of Logic*, Vol. 8, 1987.

Leonard, H.S.,

1. The Logic of Existence, *Philosophical Studies*, Vol. 7, 1956.
2. Essences, Attributes, and Predicates, *Presidential Address, 62nd Annual Meeting of the Western Division of the American Philosophical Association*, Milwaukee, 1964.

Lewis, C.S.,

1. The Institutional Use of The, *Philosophy and Phenomenological Research*, 1949.

Lindström, P.,

1. On Extensions of Elementary Logic, *Theorea*, Vol. 15, 1969.
2. A Characterization of Elementary Logic, (in) *Modality, Morality, and Other Problems of Sense and Nonsense: Essays Dedicated to Soren Hallden*, Lund, 1973.
3. On Characterization of Elementary Logic, (in) *Logical Theory and Semantic Analysis: Essays Dedicated to Stig Kanger on his 50th Birthday*, Synthese Library, Vol. 63, D.Reidel Publishing Co., Dordrecht, 1973.

Linsky, L.,
1. *Referring*, Humanities Press, New York, 1967.

Meinong, A.,
1. *Gesamtuasgabe*, (eds) Haller, R., and Kindinger, R., Graz, 1968-78.
2. Untersuchungen sur Gegenstand Theories und Psychologie (1904), translated by Chisolm, R.M., (in) his ed. *Realism and the Background of Phenomenology*, New York, 1960.

Menger, K.,
1. Wittgenstein on Formulas and Variables (in his own) *Selected Papers in Logic, Didactics, Economics, and Foundations*, D.Reidel Publishing Co., Dordrecht, 1970.

Meyer, R., Bencivenga, E., and Lambert, K.,
1. The Ineliminability of E! in Free Quantification Theory Without Identity, *Journal of Philosophical Logic*, Vol. 11, 1982.

Meyer, R., Lambert, K.,
1. Universally Free Logic and Standard Quantification Theory, *The Journal of Symbolic Logic*, Vol. 33, 1968.

Myhill, J.,
1. The Undefinability of the Set of Natural Numbers in the Ramified Principia, (in) *Bertrand Russell's Philosophy*, (ed) Nakhanikian, G., Routledge and Kegan Paul, 1974.

Parsons, T.,
1. *Non-existent Objects*, Yale University Press, New Haven, 1982.

De Pouvere, K.L.,
1. *A Method in Proofs of Undecidability with Applications to Functions in the Arithmetic of Natural Numbers*, North-Holland Publishing Co., Amsterdam, 1959.

Quine, W.V.O.,
1. *Methods of Logic*, Holt, New York, 1950.

Ramsey, P.F.,

1. *The Foundations of Mathematics*, Routledge and Kegan Paul, London, first edition edited by Braithwaite, B., 1931, and second edition edited by Mellor, D.M., under the title *Foundations - Essays in Philosophy, Logic, Mathematics and Economics*, 1979.

Rao, A.P.,

1. *Aristotle's First Philosophy in Proper Perspective*, Visva-Bharati, Santiniketan, 1969.
2. *Quine's Criterion of Ontological Commitment*, Indian Institute of Advanced Study, Shimla, 1971.
3. Towards a Free Description Theory, *Notre Dame Journal of Formal Logic*, Vol. 15, 1975.
4. A More Natural Alternative to Mostowski's Free quantification Theory, *Zeitschrift fur Mathematische Logik und Grundlagen der Mathematik*, Vol. 23, 1977.
5. On 3.203 of the Tractatus, Part Two of this volume.
6. A Survey of Free Logic, *Modern Logic*, 1996.
7. On the Indiscernibility of Facts and Propositions, *Indian Philosophical Quarterly*, Vol. 3, 1979.
8. Reflections on a Crucial Point in Leibnitz's Metaphysics, (forthcoming)

Robins, B.L.,

1. *The Definite Article in English Transformations*, The Hague, 1968.

Robinson, A.,

1. On Constrained Denotation, (in his own) *Selected Papers*, Vol. 2, (eds) Keisler, G., et alia, Yale University Press, New Haven, 1979.

Russell, B.,

1. Meinong's Theory of Complexes and Assumptions, *Mind*, 1904.
2. On Denoting, *Mind*, 1905.
3. *Principia Mathematica*, Vol. 1, Cambridge University Press, first edition 1910, and second edition 1925.
4. *Introduction to Mathematical Philosophy*, Allen and Unwin, London, 1919.
5. Introduction to Wittgenstein's (1), 1922.
6. Theory of Types, (in) *Contemporary British Philosophy*, 1st Series, (ed) Muirhead, J., Allen and Unwin, London, 1924.
7. *Our Knowledge of the External World*, second edition, W.W. Nortan, New York, 1929.

8. *Principles of Mathematics*, Cambridge University Press, 1903; second edition Allen and Unwin, London, 1947.
9. *History of Western Philosophy*, Allen and Unwin, London, 1947.
10. The Philosophy of Logical Atomism, Monist, 1918, included in his own *Logic and Knowledge*, (ed) Marsh, R., Allen and Unwin, London, 1956.
11. Mr. Strawson on Referring, *Mind*, 1957.
12. *My Philosophical Development*, Allen and Unwin, London,
13. *The Autobiography of Bertrand Russell*, Vol. 2, Allen and Unwin, Lodon, 1967.
14. On the Substitutional Theory of Classes and Relations, posthumously published in *Essays in Analysis*, (ed) Lackey, D., Allen and Unwin, London, 1973.

Rusza, I.,
1. *Modal Logic with Descriptions*, Academiai Kiado, Budapest, and Martinus Nijhaff, The Hague, 1981.

Sainsbury, M.,
1. *Russell*, Routledge and Kegan Paul, London, 1979.

Slater, B.H.,
1. Epsilon Identity, *Manuscrito*, 1993.

Scales, R.,
1. *Attribution and Reference*, University Microfilms, Ann Arbor, 1969.

Schock, R.,
1. Contributions to Syntax, Semantics, and Philosophy of Science, *Notre Dame Journal of formal Logic*, Vol. 5, 1964.

Scott, D.,
1. Existence and Description in Formal Logic, (in) *Bertrand Russell: Philosopher of the Century*, (ed) Schoenman, R., Allen and Unwin, London, 1967.
2. Identity and Existences in Intuitionist Logic, (in) *Applications of Sheaves*, Lecture Notes in Mathematics, No. 753, (ed) Fourman, M.D., Springer Verlag, Berlin, 1979.

Smiley, T.J.,

1. Propositional Functions, *Aristotelian Society Proceedings*, Suppl. Vol. 34, 1960.

Smullyam, A.F.,

1. Modality and Description, *The Journal of Symbolic Logic*, Vol. 13, 1950.

Strawson, P.F.,

1. On Referring, *Mind*, 1950.
2. *Introduction to Logical Theory*, Methuen, London, 1957.

Suppes, P.,

1. *Introduction to Logic*, van Nostard, Princeton, 1957.

Sylvan (alias Routley), R.,

1. *Exploring Meinong's Jungle and Beyond*, Research School in Social Sciences, Australian National University, Canberra, 1980.

Tharp, L.,

1. Tharp memorial issue, *Synthese*, 1989.

Wittgenstein, L.,

1. *Tractatus Logico-Philosophicus*, (1922), second translation by Pears, D., and MaCguiness, B., Routledge and Kegan Paul, London, 1961.
2. *Philosophical Investigations*, Blackwell, Oxford, 1953.
3. *Note Books 1914-1916*, Blackwell, Oxford, 1961.
4. *Remarks on the Foundations of Mathematics*, Blackwell, Oxford, 1978.
5. *Ludwig Wittgenstein: Letters to C.K. Ogden and an Appendix of Letters by Frank Plumpton Ramsey*, Routledge and Kegan Paul, London, 1973.
6. *Culture and Value*, Blackwell, Oxford, 1980.

Woodruff, P.,

1. On Super-Valuations in Free Logic, *The Journal of Symbolic Logic*, Vol. 49, 1984.

Yotsukura, S.,

1. *The Articles in English: A Structural Analysis of Usage*, The Hague, 1970.

PART TWO:

On 3.203 of the Tractatus

Chapter One : On Naming

There is nothing to be happy about the interpretations and translations of 3.203 of the Tractatus. After all these seventy seven years since it has been first translated into English, and after a whole library has been written by the interpreters of and the commentators on it, it seems that what Wittgenstein tried to encode into that remark is still eluding us. In the original it reads as follows:

> "Der Name bedeutet den gegenstand. Der Gegenstand ist seine Bedeutung. ("A" ist dasselbe zeichen wie "A''.)

For almost thirty five years by now[1] I have been reading this, and also have been without success impressing on others to read it as:

> "A name refers to an object. The object is its referent. ('A' is the same sign as 'A'.)

and not as it figures in either of the translations into English. (The difference between the two translations in so far as this remark is concerned is exhausted in the use of an indefinite article in one, and the definite article in the other.) The second translation of 1961 puts that remark into English as:

> "A name means an object. The object is its meaning ..."

This is likely to mislead the reader and induce him to believe that Wittgenstein was suggesting that meanings are entities, this is to say that he was a Meinongian of a sort. But this is contrary to the spirit of the Tractatus, and contrary to the intentions of Wittgenstein. There he was trying to work out a semantic theory which is free from Meinongian affiliations which, as he thought, can be found in the writings of Russell from the Principles to the Principia. It is well known that each of the semantic notions which Russell rejected during this period, and rejected imputing it either to Meinong or to Frege, is indeed the one which he himself subscribed to at sometime or the other at an earlier date. As dramatic irony would have it, at one time during this period, Meinong himself thought that Russell was the best representative, if not the spokesman, of his thought that he ever had in the English speaking world. On the other hand Wittgenstein was not only skeptical about Russellian semantics, but also of Fregean theories which he thought indulge in reification and multiplication of semantic categories like "proposition" and "concept". His continuous struggle from his encounter with the Principles to his completion of the Tractatus was towards a semantic theory which does not succumb to the obvious unpleasant traits of the Fregean and Russellian theories. As such he had to take a radical departure from the Fregean style of doing semantics; and he did it using the conceptual battery of Frege, however[2].

To work out a semantic theory means the same as to formulate first a criterion by means of which we can explain how a syntactically well-formed expression, that is a sentence or a word, has a "meaning", and how to elicit that meaning. Secondly, it is to formulate a criterion which would enable us to specify the conditions under which the expression that has a semantic property, for instance being true or being false, comes to have that property. This is to say that a semantic theory is geared to provide a criterion of meaning, and a criterion of truth-value, of course, in addition to providing a link between these two criteria. As the latter criterion and the criterion which has been mentioned parenthetically fall beyond the scope of this enquiry I will not touch upon them.

In the foregoing paragraph "meaning" was kept under double quotes not only to indicate that it has been a source of confusion, but also to suggest something else which may not be all that evident. So let me spell it out in some detail. There is a point in saying that in asking that the meaning of a sentence S was, we are not raising an issue in a semantic theory. The corresponding interrogative does not figure within a semantic theory because it is in response to that and similar interrogatives that a semantic theory comes into being. Of course, within the semantic theory there will be an indicative sentence correlatable to that interrogative and which can be treated as a plausible or an implausible answer to it. Such a sentence may take either of the following shapes:

(1) The meaning of S is ...

or

(2) S's meaning is ...

where S is used autonymously, as in this case itself. It is highly desirable to formulate one's semantic theory such that the very same comes out as,

(3) S's ϕ is ...

where S again is used autonymously and where ϕ is an expression of the semantic theory but is not equiform with 'meaning'. On the other hand if one casts his semantic theory such that it contains expressions of the form specified in (1) above, then the least one can and need to do is - to preclude confusions, that is - to differentiate explicitly 'meaning' as it occurs in the interrogatives of which the expressions of the semantic theory are presumed to be answers. Occurrences of the latter sort encapsulate our pre-systemic and intuitive meaning of 'meaning', and occurrences of the other sort are intended to capture the systemic and constructive meaning of 'meaning'. So to say 'meaning' in its occurrence here is as much equivocal as it is equiform.

In expressions of the form (1) and (2) above, this is to say that in its systemic and constructive use, 'meaning' does not mean any more than what can be specified in terms of that which goes into the blanks in them; and, as it should be evident from what follows, this may not be the case with pre-systemic and intuitive uses of 'meaning'. To bring this difference out to the fore, we may, in our philosophical talk about 'meaning', use two different expressions, say 'C-meaning', and 'I-meaning'. The harbinger of semantic analysis Frege[3] did note this difference. Though he noted it, he did not use two different expressions reserving one for I-meaning and the other for C-meaning; this is the case with most of the crucial passages in his writings. The cardinality of this semantic sin is not due to his inadvertency (which is not his virtue) but because it is intrinsic to the language he was using as his final metalanguage, namely German. Yet he precluded several confusions by resorting to careful and elaborate explanations about the way he intended to use his theoretical expressions in his semantic system. 'Sinn' as it occurs in ordinary German may be translatable into English as 'meaning', but as it occurs in Frege's logico-philosophical writings it cannot be translated as 'meaning' until and unless some effort is made to systematize and delineate the use of 'meaning' in this context setting it apart from its use in common parlance and make it capture the technical sense which Frege intended to give to 'Sinn'[4]. But such a task is needlessly torturous to anyone trying to construct a semantic theory using English as his ultimate metalanguage.

The central feature of Frege's semantic analysis is that the universal class of expressions is semantically homogeneous in the sense that each expression has a sense[5] and (hence) a referent. Thus his twin-function semantic commits a sentence to express a sense and refer to a truth-value. Further, within the frame of the Fregean semantics, just as the sense of a sentence is the proposition it expresses[6], the sense of a name is the concept it expresses. These can be characterized, strictly speaking named as 'that S' and 'of being N', where S and N are respectively a sentence and a name. This way of handling semantic matters, apart from the intractable problems it poses in linking the sense of an expression with its referent or the intriguing manner in which it treats epistemic contexts, does issue either in infinite regress or in reification.

Ignoring names for a while, let us take two sentences S and S', one belonging to say English and the other to some other language in which we have fair enough proficiency at least to the extent of being able to consider whether, and also decide that, the sentence S' of that language is a correct translation of the English sentence S. An explanation for the synonymy of S and S' - or to speak pre-systemically, accounting for their having the same meaning - is to be carried out in terms of sentences expressing the same sense. This sense can be referred to either by the nominalised version of S, this is to say as 'that S', or by the nominalised version of S', say nS' where n is the required nominalising functor of that language. Intuitively we know that 'that S' and 'nS'' are the same in their meaning. But there remains the issue, namely how are we to account for this intuitive knowledge within the framework of the Fregean semantics. Further, how are we to account for it without resorting to an invocation of intangible entities or resorting to a third world, or even reconciling to an infinite regress! Accounting for the intuitive knowledge of the synonymy of S and S' in terms of the identity of the senses of their nominalised versions will issue in another instance of the same problem; and accounting for it in terms of the identity of the referents of their nominalised versions results in entification and multiplication.

The formal ontology and the semantic analysis of the Tractatus are consciously designed as alternatives to the Fregean semantics and the ontology which is presumed to be either presupposed by it or seen to be a consequence of it. Whether this was done in full view of the difficulties involved in the Fregean framework or only with a partial vision of it, is not the issue here; there exists no first person account of Wittgenstein's philosophical evolution, and there is not much material constituting sufficient evidence so that a third person account of it could be given to the satisfaction of everyone. But in what Wittgenstein left to prosperity, specially in the Tractatus, there is more than sufficient evidence to confirm what I remarked about the design behind his attempt at a fresh semantic analysis[7]. Contra Frege, Wittgenstein held that the class of expressions is not homogeneous; on the other hand, as he saw it, that class is composed of two mutually exclusive and jointly exhaustive subclasses of expressions such that each of those subclasses is semantically homogeneous.

The universal class of expressions obviously is determined by the rules of syntax, and not so obviously by syntactical rules alone. (Of course in the light of what Wittgenstein came to believe a quarter of century later the latter part of this remark may stand in need of a revision; but I will not attempt at it, as here I am primarily concerned with his views of the Tractatus period.) There is no semantic role which without exception all the members of the universal class in question are expected to, or do, perform. Only sentences have senses, as 3.3 of the Tractatus would put it; and only names have referents. Sentences do not have referents, and names do not have senses[8]. I will try to offer a minimal clarification as to how this presumably obtains. Names are primitive signs; they are the basic referring expressions. This means that the possibility of a linguistic representation of the world is contingent upon names having direct reference to the objects which form the substance of the world. In any recursive characterization of the referential mechanism of language, the referring function of names will have to figure as the initial step and not either in or as the induction step. Hence

> 3.26 A name cannot be dissected any further by means of a definition[9].

Suppose that names are not that basic, this is to say that the primacy of their referring is not so fundamental in the sense that it does not constitute so to say the "transcendental" condition for the possibility of linguistic representation as such. Then it would be possible for names to have senses. Only then they can be defined in terms of other expressions[10]. If names are so definable then their reference is indirect and is derivative of the direct reference of the expressions occurring in their respective definitions. But this is not what obtains with our everyday language[11]. That it does not obtain so in our day to day language is a sufficient condition for holding that what we assumed does not constitute the "transcendental" condition for linguistic representation; this is so because such a condition must also be, at the same time, the condition satisfying which our everyday language does represent the world. In fact such a condition should be constitutive of our language. But, and that seems to be the spirit of Wittgenstein's argument, it is not a constitutive condition of our language. Despite the fact that

> 4.002 (t)he conventions on which the understanding of everyday language depends are enormously complicated,

this much seems to be certain, namely that names refer directly is one of those conventions; so must be the case with the convention that all other kinds of referring expressions, definite descriptions for instance, refer only indirectly by being related to names in ways that we can and do understand. This is to say that their reference can be elucidated in terms of the reference of names[12].

This does not, however, mean that we cannot think of the other way round being the case with language, some language that is. We can; only that would not be our language. Further, in imagining the semantic structure of such a language, or in envisioning the way we could possibly map such a language onto our language - surely, such a mapping is needed for our being able to make sense out of any expression of that language - we will be forced to a point where we may have to semantically restructure our language. That point may turn out to be one where we may have to restructure our experience itself[13]! This is likely to be so because the conventions which constitute our language enabling it to represent the world not only give 'meaning' to discourse but also contain the possibility of representing the world. These conventions are, in a sense, rules of discovery too; they enable us to discover something about the world. Thus in imagining a language in which names have secondary referring role, we will be imagining almost a new "epistemological form of life". And in such an exercise, however tempting it might be to indulge in, our received epistemological form of life and our everyday language will have no role to play. So to say, in such an activity we will be bracketing our form of life and language, and consigning them to a causally inefficacious blackhole like background. Thus in thinking of a new language which is radically different from our language, even if it were to be an ideally ideal one, we will not be attempting to understand the referential mechanism of our language; on the contrary we will be forcing it to be idle[14].

Even if the language so imagined is treated as an ideal one, it does not follow that it is a paradigmatic one from the point of view of which we need to evaluate, let alone restructure, our language in any legitimate sense of 'evaluation' and 'restructuring'. As Wittgenstein noted in On Certainty 475, a language is good enough for communication, "even primitive means of communication", is good enough, and needs no apologia from us[15]. This is why we can warrantedly believe that each sentence of everyday language, as it is given to us, is in perfect order, or as he remarked in the Tractatus

> 5.55563 In fact, all sentences of our language, just as they stand, are in perfect logical order. That utterly simple thing, which we have to formulate here, is not an image of the truth, but truth itself in its eternity. (Our problems are not abstract, but perhaps the most concrete that are there.)

To avert drifting too far let me return to the point at issue. In our language we do fling names onto objects directly. This is a brute fact of our concrete linguistic life. Thus a name comes, on its own, to refer to the thing onto which it was flung (3.261). If so, a name refers to the object to which it does refer "independently", this is to say that without being dependent upon some other name or an expression of a different kind referring to some other or the very same object. Names are autonomous in their semantic behaviour. Of course, we may fail to note what indeed the referent of a name is; we may fail to notice or understand the convention[16] which marries off monogamously a specific name with a particular object whence an elucidation will be in order. But we should recognize the fact that no such elucidation will be a definition because such an elucidation will contain the name concerning which the elucidation is being offered (3.263). 'This is N' contains 'N' in a more important sense than 'Wittgenstein' contains 'Wit'. And thinking like Wittgenstein, as demanded by him, let us note that 'Wittgenstein' does not contain 'Wit' in any relevant sense of the term. Seccability is not a virtue of names, as much as it is not a virtue of objects. Objects do not have parts; to use the scholastic terminology, they are not homoemereological. 'This is N' contains 'N' in the sense that an understanding of the former complex requires an understanding of the latter component. Standing before somebody, and perhaps pointing my finger at him, if I were to use my vocal organs to emit certain patterned sounds which, using the orthography of the English language, can be inscribed as 'This is N', and if you recognize that in this I have performed a speech-act, then it follows that you have understood the sentence figuring in my speech-act. This implies that, and that seems to be what Wittgenstein is hammering at, you have understood that 'N' refers to the biped whom I am pointing at[17]. This is the import of 3.263. A sentence can be understood only if the referents of the primitive signs which occur in it "are already known".

A name's referring to an object, and getting to know what a name refers to are different matters which should not be conflated. This should be so because only expressions which have sense can be understood; what is understood is sense. So getting to know what a name refers to requires sentential contexts because only sentences have sense (3.33). And only in a sentential context (nexus) can a name's referring to what it refers to can be either fixed or elucidated[18]. But what is the sense in which sentences have sense? What, after all, is sense! What a sentence represents is its sense, says

Wittgenstein (2.221). But what a sentence represents is a fact; that too is what he says. Are then the sense of a sentence and the fact it represents the same?[19] If our answer is in the negative we will have disturbed a big beehive of ontological issues or else will have paved the way for the Fregean type of semantics. In his attempt to refrain from doing anything which goes towards either of these, Wittgenstein is not identifying facts and senses. The world is not to be semanticized, so to say; it is to be left as it was made available to us (by fate!) without entangling it in our willing, wishing, or even "meaning endowing" - in fact free from every design of subjectivists, romanticists, and phenomenologists. Also, senses should be left to themselves without reifying; they should be left homeless without positing an autonomous world to accommodate them. Thus what needs to be recognized, as an alternative, is that

> 4.031 Instead of 'This sentence has such and such a sense', we can simply say 'This sentence represents such and such a situation'.

In this Tractatus - light, we can restate what we stated earlier. Instead of saying that what we understand when we understand a sentence is its sense, we can as well say, equivalently, that when we understand a sentence what we understand is that it represents such and such situation[20]. Further, it is important to note, the fact that a sentence represents is not its referent; facts are not named by sentences because they are not objects, and only objects are named (3.144 and 3.221). A such

> 4.024 To understand a sentence means to know what is the case if it is true. (One can understand it, therefore, without knowing whether it is true.)

It is understood by anyone who understands its constituents. Understanding, thus, is neither much more than, nor short of surmising what the world would look like if the sentence which is the object of understanding were to be true, or variantly, if the sentence were to agree with the world. This is the same as to say that understanding a sentence is getting to know what obtains in the world, or what would be the case if that sentence were to be true[21]. But, in order to surmise that, one needs to fix the objects which enter into the configuration constituting that case. This is the same as to say that one needs to know what the referents of the primitive signs of that sentence are. From the fact that a sentence stands in representing relation to the fact which would obtain as a case if that sentence were true, and that it stands in that relation to that fact in virtue if their shared form, it cannot be concluded that recognizing this form itself is understanding that sentence. This form, which the sentence itself shows forth, is shared by that sentence in common with the fact that would obtain as a case if it were to be true and also all other facts which have that form. After all, if a theory `T` takes `M` as its model, it takes every `M'` which is isomorphic with `M` as its model. Obviously, a sentence may not be uniquely exemplifying the form which it does exemplify[22]. Yet, if we understand a sentence to be standing in representing relation to that and only to that unique fact which would turn out to be the case when the sentence is true, it is because of the uniqueness of reference which the primitive signs, - names, that is - which go into the composition of that sentence do have; they uniquely refer to objects which are configured in that fact. Thus for understanding a sentence grasping its form is necessary but is not sufficient.

Perhaps, in the foregoing I packed much too much; so let me try to unpack at least two points which have direct bearing on the issue at hand which is about the way in which 3.203 needs to be understood. One of these is related to the dispensability of senses, and the other is concerned with the non-arbitrariness of the relation between a name and its referent. It may be pointed out that senses, concepts, propositions and their kith and kin are required for being able to account for the sameness of the meaning of sentences presumed to be synonymous. But inducting so large a conceptual army for solving the problem of synonymy seems to be rather imprudent. To show this let `S` and `S+` be two sentences of a language `L` of which there is intuitive certainty that they mean the same; let these sentences be say 'Socrates is taller than Simmias' and 'Simmias is shorter than Socrates'. Let `S'` be a sentence of some other language say `L'` of which again we are intuitively certain that it means the same as `S` and `S+`; let `S'` be the Hindi sentence 'Sukrat Simmias se lamba hai'. The upholders of senses will argue, as they did argue, that `S` and `S+` are synonymous in one way, and `S` and `S'` are synonymous in another way. They might say that the first pair have the same meaning because though they express different propositions, or senses, they have the same truth-value, and that the latter pair have the same meaning because they express the same proposition and also (because of it) have the same truth-value. This implies that such an explanation of the synonymy of `S`, `S+`, and `S'` results in a multiplication of the types of synonymy. That apart, it confuses, perhaps it is based upon a confusion of, signs and symbols. `S`, `S+`, and `S'` are, as Wittgenstein would say, the same sentence; the difference between them is not in their symbolic - i.e. representational - aspect but in their being different sentential signs. In the Tractatus frame of things.

> 3.32 A sign is what can be perceived of a symbol.

As distinct sentential signs, `S`, `S+`, and `S'` are the results of three different styles or modes of sign-production. (Let us juxtapose before our "mind's eyes" the Arabic numeral for three, the Roman numeral for three, and the Chinese numeral for three; do they stand for three different threes!) The apparent differences between them is inessential to their being the same symbol - being the same sentence, that is. These inessential differences can be accounted for by taking into account such empirical facts and factual factors like script, orthography, and particular grammars. But in their essence, this is to say that in so far as they are symbols, they are the same; further, it is only in so far as they are symbols that they stand in a representing relation to the world.

> 3.34 A sentence possesses essential and accidental features. Accidental features are those that result from the particular way in which the sentential sign is produced. Essential features are those without which the sentence could not express its sense.

> 3.341 So what is essential in a sentence is what all sentences that express the same sense have in common.

Then, in view of the brief elucidation on the notion of sense offered in the foregoing, it is easy to see that what is essential in a sentential sign is what all sentential signs which represent the same fact have in common - all sentential signs including those that are hieroglyphic or quasi-hieroglyphic, and quasi-orthographic ones like

`SOsi`

where the first two letters of the Roman alphabet constitute the name of the individual Socrates, and the next two letters constitute the name of the individual Simmias. We can then account for the sameness of meaning which we do frequently encounter in our everyday linguistic life by invoking and deploying the distinction between sign and symbol on the one hand, and the mode of sign-production and the mode of representation on the other.

However, this may not solve the problem as the very same issue can be raised at a different level. As it does not seem to have been raised till now in the available literature on Wittgenstein, I will try to raise it and also suggest a plausible solution to it within the framework of the semantics of the Tractatus. Suppose that we encounter a sentence `S` in some language which is not "our" language, and yet because of our proficiency in that language we do feel intuitively certain that the sentence `S'` of our language and `S` of that language mean the same[23]. Now how are we to account for this sameness in our semantic theory, which, for argument sake, can be assumed to be the same as that of the Tractatus! For instance, let S be a sentence of say the canonical Quinish[24], to wit:

> For some x, for some y : x Socratizes, y Simmisizes, and <xy> tallerises[25].

In S' and S we have two sentences, two signs which are sentences. Their discernible differences can satisfactorily be accounted for it in terms of the distinction between signs and symbols. But what about the perceptual difference? S' and S stand in two different projective relations to what they are supposed to be representing. This is to say that their modes of representation are at variance as much as two maps, one using the genithal mode of projection and other based upon the conical mode of projection stand in projectional variation[26].

The argument has now reached a much dreaded slippery slope on which many a mighty mind rolled down; so let us make sure what the issue is. There are some philosophers who are not inclined to entertain that there is a genuine issue here. For instance Quine, who has no love for propositions and concepts, but has incontestable claims for native proficiency in Quinish, may even brush away our belief that S' and S are synonymous and that this synonymy is intuitively given to us. To set us free from what he may treat as our epistemic self-deception, he may suggest his standard drug, namely the radical untranslatability thesis. Though there is nothing in the writings of Wittgenstein that suggests that for him too our belief is epistemologically deceptive, some of his followers interpreting his remarks to their advantage may set aside our belief considering it rather illegitimate, and invoke in their support a much quoted and much more misunderstood remark from the Tractatus, namely 5.62, a part of which reads as:

> "Das die welt m e i n e welt ist, das zeit sich darin, das die Grenzen der sprache (der sprache, die allein ich vertche) die Grenzen mainer welt bedueten."

If those followers of Wittgenstein fortunately have not been tempted to indulge in that sort of a linguistic solipsism it could perhaps be due to a misreading or misunderstanding of that remark, specially the parenthetical part of it. The limits of my (your, our) language refer (in the non-technical sense of the term) to the limits of my world. Their limits may coincide, all right. But which is my language? Is it "that language which alone I understand", or is it "the language which alone I understand"? That this puzzlement is not being used for rhetorical purposes should be evident because besides language there is nothing to be understood, in the strict sense of the term. When Wittgenstein wondered as to how much of the truth was there in linguistic solipsism, he was not lamenting that we (alas!) have been condemned to collective subjectivism. That this is our destiny is the belief of Quine and his followers, not of Wittgenstein though not of his followers. From his classic argument against private languages it should be evident that it is a part of his belief about linguistic phenomenon that all rule governed structures are in principle understandable by all rule making and rule following creatures. As we are such creatures, we cannot but feel sorry when ethnomethodologists among his followers make a mockery of Wittgenstein's thoughts and turn him into an undergraduate student of say Garfinkel. That apart, what is the individuator of a language? It is the set of rules of projection in accordance with which that language represents the world. If so, our everyday language is not a language; it is a bunch of languages[27]. If in our talk about our everyday language we do create the impression that we seem to take it as a language, there is nothing wrong with us because we do have in our everyday language not only multiple modes of representation, thus, in fact, a multiplicity of languages, but also rules to move from one to another; this is the same as to say that our everyday language contains rules of translation too. This is why Wittgenstein was tempted to say that

> 4.002 Everyday language is a part of human organism and is no less complicated than it.

That being the case with our everyday language, i.e. because everyday language envelopes multiple languages representing the world in multiple ways, it does make sense to talk of the synonymy of two sentences with disparate projective relations to the world. In fact we do indulge in that sort of a talk in our daily life, not to speak of philosophical life. We ask - on the strength of the fact that we do photograph, cartooning, and caricaturing - how this cartoon and that caricature are what they are respectively of the same person. This implies that there can be, as there are, two sentences which represent the same fact or have the same sense, as much as there can be two maps of a country using two different modes of projection. The sentences `S'` and `S` of our example above are thus two pictures of the very same fact, and not pictures of two different facts as Quine, the philosophical part of Quine, would have us believe.

If we gained anything till now it is a modicum of understanding of the way in which the sense of `S'` is contingent upon, or is determined by, the reference of the primitive signs which occur in it. But from what transpired till now we may not be able, as it is rather insufficient, to grasp how the sense of the Quinish sentence `S'` is determined by the reference of the primitive signs which occur in it. Apparently there are no names in it; but in reality this is not the case. In saying that there are no names in it, 'names' is not being used in its logical, or in its (essential) grammatical, sense. (It may be noted that in the preceding sentence 'name' occurs as word of meta-English but not as a term of Meta-Quinish; in its occurrence there it is not an expression of the metalanguage of the essentialism that we are talking about.) The rules of empirical grammar with reference to which we can say that there are no names in S' are not a part of the rules of projection which are unique to the language in which S figures as a sentential sign. It is in virtue of these rules expressions like 'Socrates' and 'Simmias' are names; they are names not because they are nouns which are proper names, which is only a quirk of the English language, but because they are in that language, strictly speaking in a sublanguage of English, the most primitive signs in which we can recognize the most primitive symbols which are referring to objects, and by referring so enable that sublanguage to represent the world.

The upshot of all that we said till now is that in order to net the primitive signs of S' and to understand the way in which their reference determines the sense of S' we need to surmise the way in which Quinish, of which it is a sentence, represents the world. If we look at the models for Quinish we will note that the way in which the primitive signs in it refer to objects which are members of the domain of the model is essentially the same as the way in which names behave in the sublanguage of English in question. This is not to deny that the mapping of Quinish onto its model is more complex and complicated than the way the sublanguage in question maps itself onto its model; that is obvious. Despite that, they are essentially the same in so far as they reach out their models. And on that count, and from the clarifications and explications that I tried to offer, it should follow that their models too are the same. Each of those sentences, then, reach out one and the same model - as Wittgenstein would say, "reach upto reality" - on different paths.

Earlier I remarked that when only their essential features are taken into account the names of a language turn out to be those signs whose reference figures in the initial, and not either in or as an inductive, step of a recursive characterization of the referential and representational aspects of that language. It is true that in a recursive characterization of the mapping of Quinish onto the world the initial step takes care of the variables only; and they cannot be, without being partial to the issue, considered as names because their reference does not have even a semblance of uniqueness[29]. Granting this much, let us consider the class of expressions which are truth-bearers in Quinish. What does this class consist of? It consists of, obviously, universally closed expressions containing no free variables. This is to say that it is only the closed expressions that stand either in agreement or disagreement with the world (model); these alone are sentences in Quinish. Thus, speaking in the terminology of the Tractatus, we can say that only closed expressions can have sense, for to have sense is to be in agreement or disagreement with the world. What are the referring components of a closed expression, or variantly, what are the semantically primitive terms which occur in a closed expression? The answer is obvious, namely that they are the bound variables that figure in it. Compare now the semantic behaviour of a bound variable in a Quinish sentence with the semantic behaviour of 'Socrates' in the English sentence 'Socrates is taller than Simmias', or compare it with the behaviour of 'SO' and 'si' in the quasi-hieroglyphic sentence 'SOsi'.

One can notice here too several brain-teasers. But I will attend to only two of them. First, and in the inverse of importance, the relationship between a name and the object it names is non-contingent in the sense that any use of that name requires strict compliance to the rules mapping the language in which it is a name onto the world. Of course, there is no compelling reason for calling the notorious Athenian bully 'Socrates'[30]. But when once Socrates has been given the name 'Socrates', it becomes his name; he alone is, and he alone can be, referred to by that name. This is one of the rules of representation as such that finds alternative formulations - whether these are attended to or not - in disparate languages. Language could not represent the world if names were not referentially stable - did not have constancy of reference, that is. This is why when once "a notation is established", and a mode of projection is settled for, there is nothing arbitrary or contingent in the way an object is referred to;

correspondingly there is nothing contingent in a name's referring to an object[31]. A name and its referent will have to go like Juno's swans. This brings me to the second point, which is more important in the present context. Keeping in view the referential stability of names let us try to think of the way in which objects of the model are assigned to the existentially bound variables; and in this let us concentrate on the fact that this assignment needs to be kept stable in the entire discourse (an argument, for instance). Variantly, let us think of the restrictions placed on the rule concerning the elimination of existential quantifiers, the EE as it is usually referred to. When once you use a constant to eliminate an existential quantifier you are stuck with it throughout the argument. What we notice in this, then, is that in the form of the sentence whose sense it determines, the primitive sign "will be constant" and "everything else variable". This as I see it is the insight which is well indicated in the Tractatus. The primitive sign in the case at hand happens to be a bound variable; but this is not of much consequence in understanding what names indeed are[32]. Logically speaking only such signs are names; thus it is in the essence of language that it requires names - only, in one language they will be what we call singular terms, and in another what we isolate as bound variables.

Yet, if we find some difference in the respective ways in which names like 'Socrates' and 'Simmias' on the one hand, and bound variables on the other, figure in the corresponding recursive characterizations of the reference mechanisms of the languages in which S' and S occur as sentences, it is because we have been misled by the accidental features which crept into the characterizations in question themselves. It is evident that in the characterization of the referential mechanism of Quinish, bound variables will figure in the initial step, and in the corresponding characterization concerning English, singular terms figure in the initial step. But this difference can be allowed to play a significant role in our attempt to understand what names are only by being either oblivious of or indifferent to the crucial distinction between what is essential and what is accidental in a semantic theory itself. Just because the production of signs is required for representation, for a symbol cannot be without a sign[33], the rules of representations should not be construed as if they include the rules of sign-production also. Script, orthography, etc., are not parts of grammar; similarly, in the strict sense, semantics for Quinish begins where bound variables receive referents in the model. Mapping bound variables onto the model should be treated as the initial step in working out the semantics of Quinish; and whatever is required to specify this step needs to be treated as a part of the (elaborate) specification of its (equally elaborate) mode of sign-production. This is reflective of the complexity of Quinish and its mode if representation.

We are now in a position to have a glimpse of the nature of names. A name (in essence) is a primitive symbol that functions as a vehicle of direct reference providing an initial link with the world in terms of which other referring expressions (if any) come to have reference, and whose occurrence in a sentence is required for the sentence to have first a determinate sense and next a truth-value[34]. Thus the Tractatus remark,

> 2.3411 So one could say that the real name of an object was what all symbols that signified it had in common. Thus, one by one, all kinds of composition would prove to be inessential to a name.

If that is all what being a name amounts to, names cannot have sense. All that we can say about a name is that it "refers to an object", and that "the object is its referent"; and that is all that 3.203 is saying. I have not yet attempted to explicate the import of the parenthesis in that remark, namely that 'A' is the same sign as 'A'. I shall return to it in the last chapter.

Notes

[1] See Rao (1).

[2] Engelmann in his Memoir, p.123, remarks that "when Wittgenstein went to Norway, for the first time, in 1914, to find the tranquility he needed for his concentrated mental labours, he was concerned about some remaining obscurities in the work of his first teachers, Russell and Frege, which he considered generally right and vastly superior to anything else he read. He was convinced, no doubt, that it must be possible to put those points more clearly, and the major effort on which he set out was a passionate attempt to do that". It may be noted that the central conception of the Frege-Russell view of language to which Wittgenstein subscribed at this formative period of life and philosophy was never abandoned by him. In the Tractatus he was explicit that natural language, as it is given to us, is in perfect order, but that its structure is concealed and confused by an outward clothing. It is well-known how he devoted the second phase of his intense philosophical life to discover and describe the concealed structure of this language.

[3] The sheer quantity of Frege-exegesis almost turned Frege into the Hegel of analytic philosophy.

[4] I am saying this knowing fully well that there are some students of Frege who, like Roger White, think that despite Dummett's stupendous effort at a systematization and elaboration, Frege's remarks about sense are occasional, perfunctory, and impossible to put together to form a cogent account.

[5] This is not something that can be taken lightly as it has extremely important consequences; in what follows I will be attempting to draw out some of them.

[6] This is not to be confused with the thought that the sentence manifests.

[7] Wittgenstein attempted at it after being convinced that it cannot be accomplished meaningfully; he subscribed to what has come to be called the Principles of the ineffability of semantics. His point was that the fact that a sentence represents cannot be specified without recourse to the very same sentence, and similarly the referent of a name cannot be specified without taking recourse to that name.

[8] This is a point that every reader of the Tractatus should keep in mind continuously.

[9] Subsequently Wittgenstein further strengthened this idea, of course contra Frege and Russell, to include even ostensive definition. In (10) he writes that "(i)dentifying or defining a name by ostention makes no sense, for ostention presupposes the identification of the object."

[10] See Chapter Four.

[11] What Wittgenstein was after in the Tractatus is not an ideal language and its possible link to the world; it is the essence of language in virtue of which any language whatsoever represents the world. To state the same differently he was in search of the essential language which encapsulates all the necessary and sufficient features that would enable it to represent the world. And he was not in search of it either after brushing aside our everyday language or

hoping to find a substitute for the ordinary language. His aim was to discern and isolate the mechanism in terms of which we would be able to account for the success of language in representing the world. His attitude towards everyday language can be likened to Kant's attitude towards the commonsense picture of the world. Kant did not reject that picture; his intention in the Critique of Pure Reason was, on the contrary, to attempt at a critique of it with the explicit aim of explaining how, after all, that picture does obtain. This is what propelled him to note the transcendental conditions for the possibility of experience. Similarly, Wittgenstein's view that philosophy is a "Sprachcritik" (4.0031) was geared to indicate that the business of philosophy is to discern the essential conditions of linguistic representation and explain how our language does represent in terms of those conditions.

[12] There is a more important reason for why names should have direct reference. It is that, but for it, sentences will not have unique sense. Though a sentence represents a fact in virtue of its form, only in virtue of its form it will stand in a representing relation to any (all) fact(s) which share(s) that form. Yet, when we understand a sentence S we not only surmise that it represents a fact, but also that it represents this specific fact. This specificity is the outcome of the names occurring in S having direct reference to the objects of the world. I will not explore this point further, as any move towards that end will push the discussion beyond the scope of this enquiry.

[13] It is wrong to make an epistemic or a cognitive relativist and a collective or cultural solipsist out of Wittgenstein. According to him, see his (14), concepts and conventions of other languages and cultures are neither inscrutable nor opaque. They are just useless. I know and understand how you live, only I cannot live like that; that is not my cup of tea - he seems to say.

[14] His concern for Frege's Begriffsschrift or Russell's Principia is due to his belief that they are descriptive of the structure of everyday language in its representational functioning, and not because he ever thought that they sketch an ideal language with an authority, and to which everyday language should match in order to be legitimate at all. If he tried to improve over them, it is because he thought that there were some inaccuracies in the Frege-Russell description; that is all.

[15] This is where Wittgenstein must have, like Hume earlier, experienced some agony; perhaps this agony is of the sort which only the blessed have a chance to undergo. Was not Hume distressed at the fact that the beliefs providing some harmony to our received epistemic and moral life have no rational justification! Was not he puzzled as to how the legitimacy of an argument which is cogent in itself and yet is destructive of the harmony of the received form of life? Perhaps, puzzles and distresses on these counts effected his shift from philosophy to history.

[16] Wittgenstein considered 'That is N' as a statement of a rule or a convention.

[17] What is being emphasized here is that the relationship between a name and its referent is semantically determinate but indefinable. That is why Wittgenstein was skeptical about ostensive definitions. Subsequently he suspected them on other counts too; in paragraph 28 of Part I of his (3) he remarks: "... an ostensive definition can be variously interpreted", implying that they are equivocal. Is this Quine in seminal form!

[18] In saying this I am guided by the way in which Wittgenstein, in the Tractatus, forges a semantic link between sense and reference. He construed even this link as an alternative to the way in which Frege related them in the framework of his theory, or - if one is too

demanding in giving credit - attempted theory. To clarify and compare their respective positions requires a separate enquiry; hence for the present I shall be content with noting that whereas Frege subsumes reference under sense, Wittgenstein treats sense as a function of reference, something dependent on reference. At least one of the salient features of the Wittgensteinian link between them can be elicited from the brief discussion of the concept of sense in the following pages.

[19] For some of the intriguing issues involved in this see Rao (5).

[20] As early as 22.1.1915, see his (5), Wittgenstein writes: "My whole task consists in explaining the nature of the sentence". Despairing the inability of others to see this in the Tractatus, in exasperation, as it seems, he wrote to Russell on 19.8.1919: "I am afraid you have not really got hold of my main contention to which the whole business of logical propositions is only corollary. The main point is the theory of what can be expressed by sentences, i.e. by language - (and, which comes to the same, what can be thought) ...". Whatever an explanation of the nature of the sentence might be, as Wittgenstein noticed, it will necessarily include an explanation of the expressiveness of sentences. Because of this, after Russell's misunderstanding, and rejections by Jahoda, Braumuller, Brenner, and Reclam, Wittgenstein changed the title of the projected book to "Der Satz"; at least he effected this change on the copy he had with him when he was a school-teacher at that obscure Austrian village.

[21] For an analysis of 'fact' and 'case', see Rao (1).

[22] Incidentally, if we can learn anything of first-rate importance from Plato, it is that there can be no formal theory of a particular. Such a theory will not only be a theory of a particular but also a theory of all particulars of that form. Further, as a particular is "made up" of infinite number of forms, no formal theory can be a "complete theory" of the particular, in the commonsensical sense of completeness. Thus it is the destiny of formal theories that they end up with either partial knowledge or non-unique knowledge. This may be due to the reason that no particular is unique and no particular is exhaustible; particulars may not have haecceitic properties (despite Quine); and they may have uncountably many properties!

[23] In virtue of our proficiency in that language it ceases to be alien and will be as much ours as the one in which we would hold dear to our heart.

[24] This to say that classical quantification theory with Quinean interpretation.

[25] Read <xy> as 'the ordered pair <xy>'.

[26] As we shall return to this point a little later let us just note that here is a contemporary variant of a classical problem that figured in Kant's criticism of Leibnitz. Kant's objection to Leibnitz was that there would be as many worlds as there are Leibnitzian substances. A Leibnitzian - as distinct from what one may find buried in the writings of Leibnitz - reply to this would be that there would not be as many worlds as there are substances, but only as many world pictures as there are substances. What makes these pictures, pictures of the same world is the isomorphism between any pair of those pictures. It is this isomorphism that provides the pre-established harmony which Wittgenstein obviously locates in the semantic and the syntactical structures of languages. A caricature, a cartoon, a portrait, a photograph, a statue, and a linguistic low-down of a person is a set of "somethings" about that person. These pictures are identical extensionally. They may not be so intensionally. But we can make sense of the intensional difference only when we have a definite vision of the

extensional identity. Incidentally, Quine is a Kantian in this sense of the term; and Wittgenstein is a Leibnitzian.

[27] Our linguistic behaviour is like our social behaviour; we are really polyglots - even if our linguistic knowledge is restricted to our respective mother tongues - as much as we are members of several institutions.

[28] As this specific use of this pronoun gains special significance within the Qinean world-view, it might be thought that I am having a dig at Quine. My intention is far from that, and is only towards highlighting Quine's unfair dig at Wittgenstein; see his Ontological Relativity and Other Essays where he accuses Wittgenstein of holding a "copy theory" of language when Dewey was propounding his naturalistic view. Even if this leap forward in human understanding is indicative of the superiority of the American philosophy of language over the Austrian Sprachcritik, what Quine does not seem to realize is that what he himself holds is a copy theory with two provisos, namely that a language is a copy of its world - as distinct from or opposed to the world, an expression which Quine pretends not to understand on the count of its presumed opaqueness, and that only bound variables, recently relative clauses, mirror the world. The fact of the matter is that both Wittgenstein and Quine are committed to model theory, in the broad sense of the term. In the case of Wittgenstein it was only notional as model theory developed only after he, largely under the influence of models as used in technological disciplines, formulated his Tractatus view of representation; we may recall here his knowledge of Hertz's Mechanics where entire physical theories are seen as Bilden. On the other hand, the development of Quine's views coincided with the development of model theory with which he is over familiar. The difference in their respective positions is due to the disparate ways in which they squared their respective ontological and epistemological predilections with their basic assumptions of the model theory; again I am using this term in its old fashioned and popular sense and not to refer to the discipline which is a part of mathematics. These assumptions - whether it is taken in its popular sense, or as a mathematical theory - are that individuals without inner structure are given, that properties and relations do not contribute towards individuation, and that individuation is prior to and is required for predication. But as the issues involved in these assumptions are not directly relevant to handle the problem at hand I shall not consider them further.

[29] This may be contested on the ground that the reference of free variables too is unique in the sense that Quinish can be reformulated such that $\phi x \rightarrow \phi y$ comes out, on the count of standard interpretations, invalid, or that Quinish can be reformulated using restricted quantifiers; see Chapter 4. However, conceding the maximum to the other point of view, for the present, I intend to be content with the view that at least existentially bound variables refer uniquely, and in a well-defined sense of the term.

[30] This is one of the views that Plato explores in the unfortunately unpopular dialogue Cratylus; this view was shown by him to be wrong. There is nothing in a thing which compels us to call it by the name by which we do in fact call it and not by some other name. After all, no object can suggest its own name. (Of course, there is the story that 'Plato' was suggested by its referent's having a broad chest. But being a "barbarian" I cannot comment on that).

[31] Kripke's work in the semantics of modal languages has shown the need to maintain such a stability even across-the-world representations; however, I will not be able to touch upon it here. All that I intend to point out is that the rigidity of reference of names can be noted even

with respect to representations of this world, and without invoking possible worlds even if they are tangible as this one is.

[32] 3.312 of Tractatus.

[33] A symbol is more like a universal of Aristotle rather than a Form of Plato.

[34] A name's referring to an object in its occurrences in a sentence is parasitical on and is dependent upon its having that object as its referent independent of its occurrence in a sentence. If its having a referent in its occurrence in a sentence is independent of its having an object as its referent directly, then a serious problem comes to the fore. As Wittgenstein remarked in (11) 2.0122: "what this comes to is that if it were the case that names had referents both when contained in sentences and outside them, it would, so to speak, be impossible to guarantee that in both cases they really had the same referent, in the same sense of the word. It seems impossible for words to appear in two different roles, by themselves, and in a sentence". Here we can note that Wittgenstein already had intimations of the puzzlement that identity has generated; to come out of it he is attempting to brush away another duality of Frege, specially one from his introduction to Begriffsschrift. Wittgenstein's objections to Russell's contextual elimination of definite descriptions seem to be located here; but more of these matters in the last chapter.

Chapter Two : On Natural History

It would be appropriate to begin this chapter with a characterization of the kind of the enquiry which I tried to conduct in the earlier one. If my enquiry there was Wittgensteinian, such a characterization would go a long way to help us in comprehending what Wittgenstein wanted to achieve through philosophical enquiries[1]. It is said that there are two kinds of philosophers; one of them, on better education, would turn out to be theologians, and the other if well educated, according to the saying, would turn out to be scientists. I do not think that Wittgenstein ever regretted his training, or the absence of it; in any case it is impossible for him to have received any other education than the one that he provided for himself. Consequently, the earlier enquiry about names has not been either scientific or eschatological. And if it does not qualify for the appellation 'philosophy' either, it is because, as Wittgenstein was fond of saying, it is not philosophy, but only a distant descendent of philosophy, and hence necessary.

What was I doing till now? What was I struggling to gain? Obviously I was labouring to find a plausible answer to the quadrinomial question: what is a name? I was trying to spin an explanation which would account for a symbol's being a name. This much, of course, must be evident. What, however, may not be so evident is that I intended this explanation to be in consonance with our intuitive understanding of what constitutes a name; to put it differently, I intended it to be in harmony with the understanding that each proficient user of language as such would show in using names. This means the same as that I was trying to theoretically account for a chunk of our linguistic behaviour which goes under the name of naming. Thus, it is possible to introduce a technical term at least now. I was attempting to work out a natural history of naming, and thereby a natural history of man because the former is a part of the latter.

Wittgenstein uses the term 'natural history' - in his later writings not too infrequently - in order to make important statements concerning the nature of mathematics, the nature of induction, the nature of philosophical enquiry as such, to mention a few. In the Investigations he remarks that enquiries into the nature of language are enquiries into the natural history of man; and in Remarks on the Foundations of Mathematics he holds that mathematics is not a natural history of mathematical objects. It should be obvious to any discerning reader of these and other works of Wittgenstein that he is using this expression as a technical term, a crucial one for that matter. But unfortunately, it has become a point of avoidable misunderstandings; and these, overt time, resulted in the belief - which is popular at least with a large sector of the academic readers of Wittgenstein - that he indulges in anthropology, empirical anthropology. Taking that term as if it were of common parlance - for instance as it occurs in say 'I plan to take my children to the National Museum of Natural History' - it has come to be believed that in his second innings, Wittgenstein almost took to either empirical investigations or enquiries which are similar to them. This misrepresentation of his thought could have been avoided had it been noted that there was a philosophical or technical sense in which that term has come to be used in philosophical literature for two thousand years by now[2]. It is difficult to entertain, even for argument sake, that Wittgenstein at any time in his philosophical career suspected the autonomy of philosophical enterprise, the autonomy of the kind of enquiry he was conducting.

That this has been the case with the Tractatus period of Wittgenstein's life may not be doubted by anyone; so let me cite a passage from his later writings. In 1931 he remarked:

> "A philosopher easily gets into the position of an incompetent manager who, instead of getting on with his own work and just keeping an eye on his employees to make sure they do theirs properly, takes over their work until one day he finds himself overloaded with other people's work, while his employees look on and criticizes him"[3].

Wittgenstein has always been disinclined towards disciplinary transgressions. Further, he was as much against reductions as he was against seductions[4].

Yet, if he was taken to be reducing philosophy to anthropology, it could perhaps largely be due to a large number of his readers and interpreters coming to his work during a period in which such reductions have come to dominate philosophical activity; it could even be due to their coming to his work after being acquainted with reductionist philosophy[5]. Taking 'natural history' as a term denoting an empirical enquiry concerning nature, some of his readers took Wittgenstein to be attempting at an empirical description of man; worse still is taking him to be attempting at a behavioural study of man. Characterizing his philosophical activity in this way is being rather unfair to him, because it does violence to the very spirit of his activity. Too many of his remarks in his later writings induce us to believe, without our foisting some interpretation on them, that Wittgenstein believed that empirical studies hardly contribute towards a natural history of man. To note this we should only remember his (what cannot but sound as the most) damaging comment on contemporary psychology, namely that in it there are only conceptual confusions and experimental methods. What then could be the Wittgensteinian conception of natural history!

In order to elicit what natural history according to Wittgenstein is, let us first note that philosophy, in a sense, is the natural history of man just as physics can be the natural history of matter. To gain some initial understanding let us resort to this analogy and see first what natural history of matter amounts to. An important step in our understanding of what matter was, as any history of physics would impress on us, was the development of the atomic theory. The

Greeks did introduce the idea of an atom; yet, as Wittgenstein would have said, we cannot legitimately hold that they had an atomic conception of matter. This is because their speculation (that seems to be the proper word) was so remote from our current commonsensical view to be able to provide an intelligible picture of matter; in fact that idea was far removed even from the then Greek commonsensical view so as to give them an intelligible picture of matter. Neither they saw nor we see the world as composed of the Greek atoms. What sets apart mere speculation about something from having an intelligible picture or a conceptual understanding of it can be highlighted by an illustration or two; it can be elucidated in the Wittgensteinian way - though Wittgenstein would have laughed at the way in which his style of elucidating a philosophical point is used or misused in clarifying a concept that is central to a large part of his philosophy. Suppose that someone now found

$$e=mc^2$$

inscribed twenty thousand years ago in a cave in Ethiopia! Or think of the example that Wittgenstein himself gives in his (15), namely

> "Suppose that 2000 years ago someone had invented the <u>shape</u>
>
> 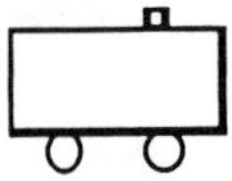
>
> and had said that one day it would be the shape of our instrument of locomotion."[6]

What are we to say of these? Obviously, we will not normally be inclined to say that the Ethiopian cave dweller was a quantum physicist, or that he had a conception of matter that envelopes mass-energy conversion; and we will not be inclined to say that because merely introducing a sign or even an idea will fall short of having a concept. Introducing signs and ideas will not be anything more that a sterile activity unless it is accompanied by something else. What is required for having a concept is, among other things, some knowledge of the way in which such an introduction helps or enables us to formulate laws of the phenomenon such that those laws are close enough to the ordinary conception of the very same phenomenon[7] (The last two words of the previous sentence have a realistic import; but I will not play on it here.) This is what Wittgenstein seems to be hinting at in the thought-experiment that he was suggesting in the lines which follow the passage quoted above.

> "Or perhaps that someone had constructed the complete mechanism of a steam engine without having any idea that or how it could be of use to anything."

Not much effort is needed to notice that to be able to ascribe an awareness of such an use, this is to say that to be able to ascribe the possession of concepts, what we need is to show that the inventor of that mechanism had all the required paraphernalia to be able to use that mechanism as a locomotive in the sense in which we use that term. But this requires a variety of things, one of which is the knowledge of the laws relating various sciences, specially mechanics. Similarly having an atomic picture of the world too requires quite a bit. To be able to have an atomic conception of the world one needs a vast amount of knowledge belonging to disparate disciplines - disparate by our standards and our conceptions. Further, the atomic conception of the world was brought closer to our ordinary conception of the world not by discovering or inventing isolated relations between these disciplines but by achieving theoretical progress culminating in a systematic theory enveloping all those and many more disciplines which hitherto remained unrelated. Such a progress is achieved by using conceptions of very extraordinary power and sophistication.

Only after such an extraordinary progress did the atomic conception of matter came closer to our ordinary conception of matter. To recount this progress which has been achieved in two thousand years in a single paragraph, it may be noted that the first step in this leap forward bringing the Greek speculation concerning matter to our conception of it (?) was the discovery of chemically pure substances allowing the distinction between atoms and molecules. With this the chemical theory developed at an accelerated pace. This development, however required only a crude idea of the atomic structure specifically the valency of particular atoms. But this could hardly tell what matter was, because it left out an enormous number of combinations of atoms which never turn up in nature at all. The second advance in the development of the atomic conception was made by relating valency to the internal structure of atoms in Rutherdford's 1913 theory, refined by quantum theory in the late twenties of this century. With this we came to have an intelligible picture of the atomic world so that we now can claim for having an atomic conception of the world. More importantly, we can claim that this conception is convincingly close to, and also accounts for, our commonsensical picture of the world. This passes the Wittgensteinian test to deserve to be called the natural history of the atomic conception of the world, or the natural history of matter.

The micro-history above was geared only to arrest our vision and induce us to look at things afresh; and keeping this in view we may ask ourselves: what is it that we are after in our attempt at a natural history? On reflection it emerges that we were relating the terms of our daily discourse to terms which go into the constitution and production of what we, in our daily life, take seriously to be the knowledge of the world. As daily discourse and daily life go hand in hand, this attempt amounts to making intelligible to ourselves how our daily life obtains; to state the same differently, it amounts to making the commonsensical conception of the world intelligible in terms of scientific conception of the world.

This distinction between the commonsensical view of the world and the scientific conception of the world should not be confused with the similar sounding distinction which Bachelard has drawn between the world as live in and the world as we think of it; I am induced to caution on that count because with the availability of Bachelard's writings in English during the last decade his distinction has become fairly popular among philosophers of the social sciences who also happen to pay attention to the later writings of Wittgenstein. For Bachelard, it may be noted, the commonsensical view of the world is devoid of cognitive content, and hence cannot be given any epistemic significance. For Wittgenstein, on the other hand, it is the commonsensical view that has cognitive content, and the scientific view of the world is parasitical on the commonsensical view. I am not saying that this is something that Wittgenstein argued out or intended to argue out. Far from it; all that I intend to point out is that the general spirit of his writings is what I am taking it to be. Surely, he was never skeptical about the scientific picture of the world; it is difficult to pick up a passage or two from his writings which can be twisted and screwed to one's advantage in order to provide even a semblance of minimal evidence to sustain that Wittgenstein brushes away current scientific knowledge; and it is possible to compile a volume consisting of remarks from his writings which goes towards establishing without doubt that he stands by the commonsensical view of the world. Thus in view of the general spirit of his work, it would not be a misrepresentation of his thought, if we say that he seems to be suggesting that say Optics would be inutile and hence indulging in it would be an idle exercise if it were not somehow hooked to, for instance, a sentence like 'The wall before me is (painted) white'. Science has cognitive content only is so far as it intersects at some point or the other with the commonsensical view of the world, and by being an explanation of how the commonsensical picture obtains. The issue of transcending the commonsensical view does not arise at all; it would be, to state it in the way he would say it, like the attempt to get out of one's own skin. We are, so to say, destined to live with the pride of our impotence to transcend it, or else live in horror of it. (This almost reminds us of the Kantian view of what enlightenment is, namely as the recognition of the limits of reason, and a reconcilement to that recognition with dignity and solemnity.)

This, however, does not mean that we cannot change our concepts or replace our intellectual batteries. Of course we can do that; and, as history shows, we have been doing it. But it appears that we change our concepts only when they are stretched far enough and their consequences are drawn out to a point where they come into conflict with our commonsensical view, where they collide with the very general qualitative aspect of our experience. Ignoring here the brute fact that our commonsensical view has over time enveloped a part of the scientific view of the world, we may note that we reject a concept on the ground that it fails to formulate laws close enough to our ordinary conception of the phenomenon, this is to say that on the ground that it falls short of providing a natural history (should we say) of man because then it develops an incongruity with "die gemeinsame menschlike Handluingdese", the behaviour of mankind (as Anscombe would put it), or the shared human way of acting (as Hintikka would say).

In the foregoing I have not unpacked the concept of natural history; what I have said till now falls short of being an explication of the concept of natural history even on philosophical standards which are not demanding. All that I offered till now may not be more than a few initial impressionistic strokes; but these are not the result of my intention to imitate Wittgenstein. Genius is inimitable; so our groping should be taken as indicative of the fact that we do not know how to go about to find a plausible explication of the concept in question. Yet, that the natural history of a concept is a part of the natural history of man is little more intelligible now, because we now are in position to surmise that, to speak the language of Kant, natural history exhibits the transcendental grounds of our commonsensical view, the view on the basis of which we live and act in our daily life. This initial understanding should go towards legitimizing natural history as a discipline because that is sufficient to turn our attempt at natural history into a worthwhile pursuit. At the same time such an understanding raises several issues, one of which may be formulated as: how much of natural history is needed to understand man and his activity. Wittgenstein seems to have thought that much of it, this is to say the natural history of many a concept, is indeed wasteful for the purposes of that understanding. Take for instance the case of the concept of language itself!

Much of the natural history of language (in the ordinary sense of the term), worked out assiduously during the period of reconnaissance and later during the high tide of colonial expansion and consolidation, has not been of much use in isolating the essential nature of human languages or human linguistic activity. This does not, however, mean that all that work which certainly is a monument of human labour does not contain any interesting ideas; only those ideas, some of which are really brilliant, are not sufficient to spin a systematic theory of language as such. One of the reasons for this insufficiency can be located in the way in which those linguists, professional or amateur, conducted their respective enquiries; they largely concentrated on the idiosyncrasies of individual languages, and compared and contrasted those peculiarities. Never they searched for what we now call "linguistic universals."[8]

Contrast this situation with what has happened, and also is happening, in genetics. (The reader may bear with my quantum-leaps in driving a point home; but that seems to be the only way in which I can go about in the present enquiry.) Restricting the area of investigation and concentrating on bacteria and virus proved helpful in developing a systematic and convincing science of genetics. Adopting this strategy one could observe in a short span of time many generations of species enabling him to note how the difference in degree could be explained to produce the difference in kind, for instance. In a similar spirit, attempting at a convincing natural history of language, specially in its representational aspect, Wittgenstein concentrates on some unique kinds of languages; these are the kinds of languages that we do use in our daily life, the kinds of languages that we would still be using even if we restricted our lives to a bare minimum and lived restricted lives. These are the kind of languages the use of which even in lives so restricted can still be considered to be integral to the very functioning of human organism itself. After all we and our languages both have evolved in the world we live in. So why do not we use our built-in views to give a reliable representation of the world as well as the languages! Why do not we use our built-in views in our self-understanding! Why should we see this as an obstacle between us and the world, or between us and our language!

At this stage we may repeat the question: how much of the natural history of language (in the colloquial sense of the term) proved to be useful in making intelligible what language is? The reply if not encouraging, of course. But contrast the situation with the attempts towards a natural history of language in the technical, this is to say that philosophical, sense of the term. Though we cannot say that we now have a convincing theory of language, variantly put, though we cannot claim to have worked out a natural history of the concept of language, we have nevertheless moved quite a long way in that direction. An important step towards the development of a theory of language was the discovery of the essential linguistic component - which for convenience we shall call "essential language" - embedded in all natural languages. This allowed the distinction between the essential and the accidental features of a language. With this discovery a theory of language - a theory of linguistic representation - came to be developed, as it required only a primitive, may be even crude, idea of sense and a similar idea of reference; this is the point where Wittgenstein's <u>Tractatus</u> figures, and also takes a lead. But this proved to be insufficient to tell us what language was. The reason for this shortcoming is obvious, namely language, specially the language we use in our day to day communication, has enormous number of expressions, simple as well as complex, which have no direct, may be not even remote, role in representation.

Then a significant advance was made by relating sense and reference to use; that this advance was due to Wittgenstein of (4) and (3) hardly needs a mention. But all this has not yielded a natural history of language; yet it is fecund enough to induce us to believe that we are now somewhere near about it - at least on the right path towards it.

Even if it is conceded that what I am pointing at enables one to understand what it is to do the natural history of a concept, several problems and puzzles still remain unattended; for instance, one might be puzzled as to what natural history, after all, is all about, because if he is familiar with Wittgenstein's writings he will have noticed that he sometimes talks about the natural history of concepts, and then moves to talk about the natural history of man. It seems to me that such a puzzlement can arise only when a brute fact of linguistic life is not taken into cognizance. Representation is so central in our linguistic life that we need not harp on the distinction between the conceptual realm and the mundane world which is its "extension". For example, saying that Wittgenstein was a philosopher, and saying that the sentence 'Wittgenstein is a philosopher' is true seem to do the same even though one is about a fact (the world) and the other is about a sentence (the language). In our linguistic life we do the same with either of these speech-acts. This is not to say that the material and the formal modes of speech and enquiry need not be differentiated; it is obvious that there are contexts where not to differentiate them would tend our enquiries towards anomalies and confusions. All that I am suggesting is that there are contexts, for instance the one mentioned above, where the two modes of speech play equi-effective roles in our linguistic life. If it were so, there may not be any feature of a fact (the world) that is independent of the features of the corresponding sentences (language). This is precisely why the picture theory of the Tractatus was intended to go into the natural history of the world as well as of language. In any case it was designed to into the natural history of man[9].

The borrowed Kantian digression which I relegated to an endnote - to preclude a break, obviously - is extremely important to understand what Wittgenstein was doing in his (12), specially his explication of the concept of a proof, and his rejection of (Brouwerian) intuitionism. His basic argument has structural similarities with the argument of Kant[10]; and the common point of these two arguments is relevant to the point of discussion here; and it is relevant in the following way. If what I said about the Ethiopian cave dweller is true, we will not be able to say that Wittgenstein gave a concept of natural history until and unless we are in a position to show how using that concept he worked out a convincing theory of at least one concept. This is the least that anyone can expect; so let me attend to that by showing how he worked out a natural history of the concept of proof.

As a word of caution, or to preclude a misunderstanding, of my venture I may state explicitly that in attempting what I am about to attempt I am not trying to be a Shorey or Stirling of Wittgenstein; nor he shared with me (in dreams, that is) what he really meant. I am not even suggesting that this is the way Wittgenstein thought, or that this is what he thought he was doing. It could be that thinking is a play, and that no thinker does write his script consciously. Thus my point is that there is no reason why we should not see Wittgenstein's contribution the way I am trying to look at it.

The Nestroy passage[11] which figures as the motto of the Investigations is more appropriate to Remarks on the Foundations of Mathematics[12]. There is something of the Galilean conception of mathematics in the Remarks view of it. By holding that mathematics is what mathematics does, Galileo pushed the ontological status of mathematical entities from the realm of philosophy to the domain of theology. What Wittgenstein is trying to do in the Remarks amounts to pushing the remnants of that Platonic legacy from the realm of theology finally to the area of folklore, the ludicrous i.e. This seems to be what he intends to gain in arguing, contra Platonists, that mathematics is not a natural history of mathematical objects[13], but only of proof. This view of mathematics is already present in the Tractatus, in its seminal form though. At 6.211 Wittgenstein remarks:

> "Indeed in real life[14] a mathematical sentence is never what we want. Rather we make use of mathematical sentences only in inferences from sentences that do not belong to mathematics. In philosophy the question "what we actually use this word or this sentence for?" repeatedly leads to valuable insights."

Let me then contrive an illustration which may exhibit how this is convincing in the sense that it appears to be closer to our intuitive conception of mathematics, or nearer to the conception of mathematics that seems to be embedded in our daily activity. Suppose you want to get a specific quantum of work done in a specific period of time by engaging the required number of workers, each of whom does some specific amount of work. To find out the number of workers you need to hire, you formulate a quadratic equation. Now suppose that this equation has two solutions, say m and $-n$. You will naturally contact a ganger and hire m number of workers. But why? You could have engaged $-n$ number of workers as well! [15]

If there is a genuine puzzlement here, it should be the result of an illegitimate belief to the effect that in mathematics there is choice; and a little reflection would show that there is no such choice. Personal preferences have no role to play say in arithmetic. Mathematical proofs, in general tell us:

> "Yes, this is how it has to be,"

and

> "This must be like this."

However, this is not due to the nature of "mathematical reality" or because of the characteristic features of the "mathematical objects". The necessity which we encounter in mathematics is not due to any of these - there is, in fact, neither a mathematical reality, nor are there mathematical objects - but due to the nature of we human beings. This simple fact, which seems to have eluded us till now, in the Wittgensteinian framework, is as much a constitutive component of human nature as is the three dimensional Euclidean representation constitutive of human mind for Kant. The concept of number that is embedded in our representation of the world[16] does not leave us any freedom to hire either "negative men" or "negative number of men". And with our settling for a concept, we are settled (Investigations, 1.230), or as Wittgenstein puts the same in Remarks, 1.119:

> "The hardness of mathematics is indeed our hardness".

We are not free from the puzzlement yet; we may still feel some uneasiness concerning another issue, namely why is that we embedded, this is to say that why is it that our language embeds, this concept of number in our representation of the world, and not some other concept which might have permitted us to hire negative number of workers! Wittgenstein has a long-winding argument to let philosophers as well as mathematicians out of this "fly bottle"; I will try to recast it in its bare essentials. (This argument has been thought out by Wittgenstein in connection with his belief that there is a point in doing mathematics in the classical way while accounting for the possibility if doing it in the intuitionist way. As a detailed consideration of this part of his contribution to the philosophy of mathematics has no direct bearing on the point under consideration, I shall reserve it for another occasion. For the purposes of this enquiry it would be sufficient to pick the central thread of that argument and state it relativising it to the issue at hand.) Earlier I noted that the concept of number that is embedded in our language is constitutive of our nature. Then to say that we could have gone for a different concept of number amounts to saying that we could have been different from what we indeed are[17]. It may be true that "the formation of concepts can be explained by the facts of nature,"

only, as Wittgenstein would say, mathematical concepts are not of that type, because mathematics does not state facts of nature (12.1.116). This idea is an integral part of Wittgenstein's hardness; he continued to subscribe to it since (5) and (1) onwards. All that mathematics does is to provide a framework for handling those facts, a framework in which we alone, and in which alone we can handle facts of nature.

Let us try, then, to imagine that we are different from what we are, this is to say that the facts of our nature are different from what they are. Then

> "the formation of concepts different from the usual ones becomes intelligible" (3.1.p.230).

If this is possible (at all) - I personally do not know what it all amounts to, or how to go about with it - this is to say that if we are not what we are, or variantly

> "(i)f the formation of concepts can be explained by facts of nature,"

then

> "should we not be interested, not in grammar, but rather in that in nature which is the basis of grammar?" (3.11.XII)

A clarification regarding this thought-experiment is in order; and fortunately Wittgenstein himself provides it. Continuing the passage quoted above he writes:

> "I am not saying if such-and-such facts of nature were different, people would have different concepts (in the sense of a hypothesis)[18]. But, if anyone believes that ... having different ones would mean not realizing something we realize - then let him imagine certain very general facts of nature to be different from what we are used to, and the formation of concepts different from the usual ones will become intelligible to <u>him</u>."

The last pronoun which I am emphasizing seems to be crucial to understand the intention of Wittgenstein because the whole irony of the passage, it appears, is encapsulated in it. If so, we may have to read it as: will become intelligible to him if he really reflects on what is involved in such an exercise and sees through what it all amounts to. Perhaps in addition to other things, it amounts to saying that we have the concept of number which we do have because we are what we are, and that if we are to be different so would our (?) concept of number be. (One needs not be semantically extra-sensitive to note the referential ambiguities involved in the different occurrences of the pronoun here; but I will not consider it, as I am not concerned with it here.)

In order to think of the ways to represent "reality" in language embedding a concept of number which is different from ours, thus in order to imagine an arithmetical representation of the world in a different arithmetic, we have to use, in the intellectual exercise in question, arithmetical rules which our arithmetic may prohibit or forbid. The rules of our arithmetic differentiate the permissible representations from the prohibited ones in a way that is unique to it, and in a fashion that is uniquely acceptable to us. In so far as we are concerned the rules of our arithmetic demand and obtain universal and constant compliance from us. This does not, however, mean that there is some essential or necessary connection between our concept of number and the world[19]. Belief in such a connection, as Wittgenstein would say, is "dangerous and deceptive"; this is so because it would lead either to a reification of concepts or else to a conceptualization of the objects. To preclude these we need to refrain from assimilating "facts of nature" and "the determination or formation of a concept". Wittgenstein cautions us on that count in the Remarks on Foundations of Mathematics. It is this sort of an assimilation turning a concept into a part of nature which Wittgenstein is attacking in Lectures on the Foundations of Mathematics; he writes on p.82

> "if you find something natural (say Euclid) then you will find other things unintelligible (say Gauss, Labacheski, Bolyai, and Riemann). Are the latter unnatural? Only if one is confusedly enmeshed in Euclid."[20]

One of the ways in which we get into this sort of an enmeshing can be elicited from the following passages from (12.11.3)

> "The dangerous, deceptive things about the idea; "The real number cannot be arranged in a series", or again, "The set is not denumerable" resides in its making what is determination, formation, of a concept look like a fact of nature."

(Wittgenstein seems to have grafted this idea from Weyl's 1921 paper in Mathematische Zeitschrift. If Kreisel reports that he is familiar with this paper, von Wright reports that in one of the type scripts of the Investigations there is a marginal insertion by the author to the paragraph 352 which reads as: "Says Weyl". As I am not interested in textual matters, refraining from eschatological digressions, I may just point out that for Weyl such expressions are not sentences, but only sentence-licenses.) Such dangerous or deceptive confusions are not unique in the case of mathematical "statements"; throughout our language many such confusions have crept in, and many more are likely to creep in. Do we not, occasionally at least, conflate the imperative and the indicative moods? Compare for instance the two sentences

> I want you to hang yourself.

and

> You should hang yourself.

If you are obliging, the effect will pleasingly be the same. However, that is not of any interest to philosophy. What is of philosophical interest is the difference between these two sentences, or the very fact that there is some difference between them. The first sentence is about me or my ill-will in a precisely definable sense of the term. But the second sentence is not; it is not about you either. It is this sort of a confusion that Wittgenstein is suggesting to be residing at the root of set theory when he says that what we notice is an admixture of methods of proof and conceptual confusions (3.11.XIV). That an ordinal is the set of all of its predecessors is not a statement about ordinals; nor is it a statement about sets. It is a rule to generate a series, a determination or formation rule, as Wittgenstein would say. Now suppose that we eschew reification of mathematical objects or ontologicising mathematical concepts, then what will be left in mathematics? Obviously, only methods of proof; and this is precisely the Galilean view of mathematics.

But what is a proof? There is too much in Wittgenstein's writings, Wittgenstein-studies, and elsewhere which needs to be taken into account in order to be able to provide a reasonably reply to this. So let me be less ambitious and ask: what is central in a proof? The formalist account that it is a transformation of a set of well-formed formulas to another set of well-formed formulas satisfying certain well-defined conditions does not throw much light on the nature of proofs; for philosophical purposes this account is inadequate. Even a practicing mathematician, with his habitual aversion to philosophy, cannot find sufficient protein in the formalist account; thus the importance of what Wittgenstein has to say on proof. A proof, as he sees it, first of all "introduces a new concept". By this he

> "meant something like: a proof puts a new paradigm of language" (12.1.31)

As paradigm are prescriptions - they demand and secure universal unforced compliance - a new proof thus issues in a new prescription. This seems to be so in two different senses, one trivial and the other rather subtle. For instance, once the deduction theorem is proved making use of the primitive rules of inference, we will have what are called the derived rules of inference such that we can say with some legitimacy that we have a new concept of a rule of inference. This is the trivial sense in which proofs are said to introduce concepts. Compare this with what has happened in the development of set theory from Cantor to Cohen. With its being kept on an axiomatic base, with the obtaining of the independence results, and with all that has happened in between, we reached a point we can ascribe to ourselves the possession of a concept of a set that is qualitatively different from the one embedded in Cantor's naive or intuitive definition, indeed characterization[22], of a set such that we can say, with some legitimacy, that we have a new concept of a set.

In the foregoing I have been talking about the concept of a set, and the concept of a number. But what after all is a concept itself? To strip it off from the mystique aura with which some philosophers seem to use 'concept' and to note its nature in its nakedness, a concept is a bunch of rules[23]. Concepts are constituted by rules. Then, when once reification is refrained from, where are these rules to be located? How are they to be grounded? These were serious problems for Frege too; having refrained from psychologicising them, he too faced those issues. But having argued why rules are not to be kept in the world, and why we should not treat them as rules of thought either, he came to hold that they are "the boundary stones fixed in an eternal soil" - an eternal soil of thoughts and theories, concepts and propositions, what Popper later were to call the Third World. This tertium mundus must have offended Wittgenstein's philosophical sensibility as much as Frege's twin-function semantics must have offended his semantic taste. So here too he took a departure from Frege and joined hands with William of Occham in not multiplying worlds. His point was that these rules are constitutive of man[24]. It is not the case that in the beginning there was man and then he created rules and decided to follow them. Man is rule-following, rule-inventing, and rule-revising being. These activities are integral to what it is to be a human being. This is implicit in the writings of the second phrase of Wittgenstein's philosophy, and is explicit from the following passages from the Investigations, 1:

> "Commanding, questioning, recounting, chatting, are as much part of our natural history, as walking, eating, drinking, playing."[25]
> "What does man think for? ... we are not interested in causes - we shall say: Human beings do in fact think." (466)

> "Does a man think because he has found thinking pays? Does he bring up his children because he has found it pays?" (467)

So is the case with counting. From my son's setting apart four toffees into two groups of two each and sharing them equally with his sister, to my hesitant flights into Cantor's Paradise or Cohen's Hell, all this is a part of the way "in which we act"[26]. What my son and I do are rule governed activities, conceptual plays, so to say. But where are the corresponding rules, the concept of number, the concept of set, etc., to be located? This is the same as to ask, where are the rules of arithmetic to be grounded? The obvious answer, and a plausible answer too, is: in my and my son's activities which are parts of the individuators, or the constitutive conditions, of human beings which I and my son happen to be[27].

If the foregoing has a grain of truth in it, we may have to face another serious problem. If the rules in question are constitutive of what we are, how can we account for conceptual changes? To ask the same differently: is Wittgenstein's thought a variety of conservatism? Let me try to answer this issue by splitting it into two sub-issues. These are: first, does what Wittgenstein is saying provide for conceptual changes and revisions? Secondly, if it does provide for that, how to account for such changes within its framework? Here I need to handle only the first of these; so I shall restrict my concern to it. Even so cautious a scholar as Walsh, whose judgments on others are well balanced[28], thought that Wittgenstein was a conservative of the Burkean sort. But as I see it this is rather unfair and may have been occasioned by Walsh's failure to differentiate between two central tenets of Burkean conservatism. These are, first that the existence of a rule, or a practice, or an institution, is the evidence of its rationality, and secondly that (hence) demanding the existing rules conform to the requirements of the imagined alternatives is irrational. Wittgenstein would subscribe to the second tenet; but not to the first, I trust[29]. Rules, practices, institutions, etc., are for him constitutive of our being, all right; but they are so only contingently. They do not constitute us essentially[30].

This is so because

> "essence is not characteristic of an object," (12.1.73)

in fact of anything which is a part of nature[31]. And

> "if you talk about essence, you are merely mentioning a convention" (ibid.74) -

a rule that is. Because rules determine concepts in the sense in which members determine a set, concepts can be said to have essences as much as sets can be said to have essences. We need, however, keep in our view that concepts are determinate, and that the rest are indeterminate and open. In fact we have been revising and changing our concepts; we have been continuously introducing new concepts.

Further, it is not legitimate to talk about the rationality of the concepts which we do in fact use because first, it is these very concepts which go into any possible explanation which we may be able to offer for something's being rational or irrational. In view of this, in so far as the second tenet of conservatism is concerned, Wittgenstein's opinions can be taken to be in agreement with those of Burke. Secondly, the issue of the rationality of rules and practices which are constitutive of our being does not arise at all because these amount to the normative structure of our activity itself. As the ultimate bases of our action are the ultimate bases of our thought too - this is something of which Wittgenstein was convinced of - the only reason for our acting in accordance with the rules in accordance with which we do in fact act is that we, after all, do act so (7.284). The search for the grounds of our thinking must come to an end sooner or later, but the

> "end is not an ungrounded presupposition; it is an ungrounded way of acting." (ibid.110)

The ungrounded way of acting is the way in which "we find it natural to do" (ibid.471) - or, as the Buddha might have said, the way in following which we are at ease with ourselves, with each other, and with the world. That apart, the final ground of our action and thought, Wittgenstein is suggesting, is a fact about us, a fact of the natural history of man[32]. This has a Kantian ring - whether Wittgenstein was conscious of it or not - for to reduce Kant's argument to its bare bones and characterize it, what Kant was attempting at in the first Critique is to show, whether he was aware of it or not, how we being what we are, know what we know in the way in which we indeed know, this is to say that our epistemological statements are statements about us[33].

To the professional or academic philosophers all this may sound rather complaisantly complacent, all the more so because there are tiringly countable number of passages in the later writings of Wittgenstein which create the impression that he is disinclined to see the problem. These passages cover a wide range of subjects from mathematics to matters that are mundane. As I intend to present an idea to a wider audience, I propose to take into account those passages to understand that no specific disciplinary knowledge is required. Consider, thus, for instance

> "(w)e have a colour-system as we have a number system. Do the systems reside in our nature or in the nature of things?" (13.357).

Let it be that these systems are in our nature, and let our nature and our practices be related in the way in which Wittgenstein finds them related. Still one can wonder whether it is not going too far to say that

> "the bases of our practices are not called in question: they are there - "like our life." (7.559).

To take into account the number system a little later, let me first consider our colour-system. We have our colour-system[34] just as, as the entomologists say, the butterflies have theirs. We represent (describe) a butterfly in our polychromatic system; and another butterfly represents the very same butterfly (spelling out the realistic bias here is not needed) in the dichromatic system of butterflies, this is to say that in a system of black and white, of course different shades of black and white. Can we and the butterflies swap our respective ways of representation? Perhaps not; such a swapping is impossible because we and the butterflies are constrained by our respective ocular apparatus and our respective cerebral mechanism[35]. Thus, Wittgenstein argues that

> "(t)ill now only the ratiocinative element in knowledge has received attention - the animal part of it too be given proper place." (7.475)

Nevertheless, physiology seems to be irrelevant here. It may not permit us to transcend our cognitive practice. But that is besides the point because what matters is not that we cannot get out of our cognitive practice, but what we need not; and we need not think of it as long as we are what we are. Even if such constraints do not exist, there does not seem to be any point in such swapping because representations of the butterflies are geared to their interests of which reproduction and food-gathering seem to be the dominant ones, and our representations are geared to our interests of which recreation is one, aesthetic enjoyment another, and scientific knowledge yet another. This is to say that our and the butterflies' cognitive interests are different. And when there are no interest-shifts, conceptual changes are hardly in order. One sees things differently if his interests are different[36].

Can we say the same thing about our number system too? There do not seem to be any physiological constraints involved in this case; and what sort of a change in our interests is required so that all of us may turn into intuitionists, and do mathematics à la Brouwer and Heyting! Why do not we see things from tomorrow morning as the intuitionists claim to see[37]! Does our becoming intuitionists amount to changing our form of life! If there is no point in such a conversion, it is because our interest remains the same - it still is a representation of the world; and there is no evidence to the effect that the intuitionist way of doing mathematics serves this cause any better. But our hardness - our tenacity with our ways of doing things - is not due to this only; it seems that we stick to our practice for better reasons as well. Wittgenstein suggested at least one such.

No mathematician actively involved in the practice ever thought that the Brouwerian bombardment would batter his bastions. No classical analyst, for instance, felt lost at not being able to provide proofs which are pleasing to the intuitionists. Working mathematicians - Brouwer too is one of them, as he thought - responded to intuitionism with the same composure which the physicists of the yore showed in dismissing Zeno's puzzles and is befitting their profession. If Wittgenstein is right - he seems to be right - in holding that a proof introduces a new concept of number, the intuitionists are introducing a new concept of number. For in constructing their proofs they refrain from using the law of excluded middle, let us note what this law and its use amount to, and then see what bearing it has on the concept of proof itself.

Crispin Wright has correctly pointed out that the intuitionists as well as Wittgenstein took this law into their cognizance in its stronger version, this is to say that understanding it to be asserting that the disjunction of a sentence and its negation is valid; in its weaker version it is assumed to assert that there is no third value, besides 'true' and 'false'. Why the intuitionists accept the stronger version is a long story which is, though interesting, of no direct relevance to the point at hand; so let me try to note the reasons for Wittgenstein's accepting it in that version and also his rejecting the weaker version. He rejected the weaker version as he thought that the tertium principle does not make sense. He wondered in his Lectures on the Foundations of Mathematics:

" 'a sentence is either true or false' - well, what is the third possibility?"

Otherwise, this is to say that if it is neither true or false, it is not a sentence at all. This, in a sense, is the definition of a sentence; it shows what the concept of a sentence is, so to say. There is, however, another reason for rejecting the weaker version. There is no textual evidence to the effect that Wittgenstein is aware of it. But the point that Aristotle raises in connection with that kind of rejecting the law is raised by Wittgenstein; in fact in his (12) he uses it in order to show how a shared concept of proof is required for settling the debate between the classicists and the intuitionists. Thus, it may not be out of place to briefly state it here. Paraphrasing Aristotle we may point that if a sentence is neither true or false, let us call it doubtful. But if the law in question is false, it needs to be either doubtful or not doubtful. Thus we shall have not just three possibilities, namely 'true', 'false', and 'doubtful', but four, namely that it is true, that it is false, that it is doubtful, and that it is neither true, nor false, nor doubtful; thus we can go on ad infinitum. But if one says why not? - this point is paid sufficient attention in (12) - then there is nothing more to be said. We cannot see how any common basis can be found from which we can deliberate about the matter. Wright hints at why the intuitionists could not have gone for the weaker version; it is because they accept that the double denial of the disjunction of a sentence and its denial is valid. But as he does not tell us why Wittgenstein does not go for this version, we may add another reason and provide a fair picture of his choice. The issue is not such about - as he came to see - there being a third possibility. It is, as he saw it, about:

> "(d)oes reality accord with this picture or not."

I believe that we should take this more seriously as it occurs in the Investigations (1.352); and the reason for my giving greater importance to this work is that Wittgenstein wanted to publish this work, though after it was rejected by the Cambridge University Press he dropped the idea (it was subsequently published posthumously) - and wrote a preface. Thus we may treat other remarks from the posthumously published writings for which the author's prefaces do not exist as personal notes towards works which the author did not intend for public presentation in the form in which they were recorded.

Wittgenstein accepted the stronger version of the law of excluded middle because accepting that version and then arguing that mathematics does not in the strict sense of the term contain statements but only rules comes to be a better strategy for this provides him, as he must have noticed, a single stick with which he could beat both the intuitionists and the Platonists. If the sentences

> "Every number greater than 2 can be represented as a sum of two primes"

and

> "There is an even number which is greater than 2 but cannot be represented as the sum of two primes"

both are statements of rules. Platonism is misleading as much as intuitionism is inutile; this is so because then one of the sentences is not contradictory of the other. "You should do x" and "You should not do x" contradict each other; but this is not the case with "You should do x" and You should do not-x". Further, how is the law of excluded middle itself to be used? There must be some prescription indicating where its use is permitted, and where its use is prohibited. In any case, what does the non-availability of a proof of either of the two sentences given above indicate? - it does not indicate that we should refrain from applying the law of excluded middle on infinite totalities; nor does it indicate that we should not use it in arguments concerning entities whose existence proofs are not constructive in the intuitionist sense of the term. But it does seem to point at the fact that, among other things, the concept of number, our number concept that is, is not yet fully determined. This is to say that what Wittgenstein thought of set theory has an analogue concerning number theory.

Here too we have conceptual fuzziness and methods of proof[40]. Nevertheless, this fuzziness - this is extremely important to understand the Wittgensteinian point of view - should in no way constrain us in using the concept of number. The failure of Hilbert's program need not paralyze us; we can still go ahead with our arithmetical representation of the world, and go ahead legitimately. In the Remarks on the Foundations of Mathematics he was emphatic about it when he wrote

> "...(t)o use a word without justification does not mean to use it without right,"

and

> "(T)he danger here, I believe, is that of giving a justification of our procedures where there is no such justification, and we ought simply to have said "that is how we do it"."

Is mathematics, then, a description of a chunk of life and activity? Wittgenstein's answer is: No. Mathematics is the method of proof; it is a set of rules constituting, so to say, the statute book of our representation of the world in language, as much as grammar is. Then,

> "...(are) sentences of mathematics anthropological sentences saying how men infer and catalogue? - is a statute book a work of anthropology telling how people of this nation deal with a thief etc.? Could it be said,"

he continues in (12),

> "the judge looks up a book about anthropology and there upon sentences the thief to a term of imprisonment? Well, the judge does not use the statute book as a manual of anthropology."

The statute book is not used as a field manual or as a map. How does the judge use it, then? Obviously, he uses it by taking it "to regulate his performance of his office." Mathematics thus is normative; encoded in it are the norms to which our representations should conform and do in fact conform. The natural history of mathematics is geared to account for this normative character.

Coda:

Now it should be easy to surmise what I meant when I remarked that in the first chapter I was attempting to work out a natural history of naming. Taking what emerged in the first and this chapters into account we even understand what it means to say that we do use names. Yet there is a rub; and it will be occasioned by Wittgenstein's remark in the Tractatus that we can represent the world without resorting to names[41]. Does it follow from this that names can be dispensed with, this is to say that we can do without using names? In the next chapter I shall argue that it does not.

Notes

[1] Whether this kind of an enquiry is philosophical at all has been debated; so we might as well drop the qualification 'philosophical' and try to understand what constitutes a Wittgensteinian enquiry.

[2] For instance, when in De Anima Aristotle set out to work on the natural history of "psyche" he was not attempting to do experimental psychology.

[3] Emphasis added.

[4] The type of philosophy which results in assertions like: x is not only x, but also y.

[5] A case in point is Quine's reduction of ontology to semantics, semantics to epistemology, and epistemology to natural science; another is Kuhn's reduction of natural science to academic politics and intra-cirricular intrigue. The appearance of much of the Wittgensteiniana coincided with he spreading of these reductionist philosophies.

[6] p.43; the passage is of 1943 vintage.

[7] What I am suggesting here might sound like a variant of the popular view that science is an extension of commonsense; but I would like to caution that this is not what I mean. Scientific theories may be counter-intuitive, and commonsense may come to change in the light of the accepted scientific theories. There seems to be a two-way relationship between these two.

[8] This task was taken up by Chomsky and his followers. But Chomsky too fails to give us a convincing theory of language. His invocation of the a priori, and his last resort naturalism (of the biological variety) fail him in convincing us that the way he sees is the way our linguistic life is.

[9] A Kantian Digression: Similarly, it is idle to be puzzled at whether in the first Critique Kant is giving us a natural history of mind or a natural history of the concept of human knowledge. This analogy itself might puzzle some, without further clarification, that is; so I borrow one. With this clarification another fact of natural history as a discipline should come to the fore.

Kant's "research program" has many important features though only two of them are of concern for us here. First, Kant did not attempt either to bypass or transcend the commonsensical picture of the world. On the contrary, his intention was to show how that picture obtains. The second feature which should interest us here is the way he executed his program. Let us note, evading Kant exegesis, what Kant was up to in the first Critique. Let us state it fusing the formal and the material mode, as he himself does; that would vindicate our point which was mentioned earlier, besides serving the intended purpose. The empiricists and the rationalists before Kant had their respective conceptions of human knowledge. Kant rejected both these conceptions on the ground that the formulation of neither of those theories is closer to the way in which we intuitively think that our minds work, or how we think that we acquire knowledge. Both those theories start with the assumption that there is a class of sentences whose truth-conditions we know we grasp in our primitive interaction with the world. So to say they are the paradigms for settling whether or not other sentences are true, even how other sentences are to be understood. We may call the former class the "preferred class of sentences", and the latter class the "problematic class of sentences". The empiricists and the rationalists differ in their respective choice of the preferred class; they go for different preferred classes. Kant's point is that the preferred class as an input will not give us the problematic class as the output unless we have at our disposal a program which neither the empiricist program nor the rationalist program either provided or allowed for. In order to write such a program we need to understand what the mind must be like so that it may make it possible to arrive at the intended output from the given input. This program, this is to say that a theory of mind, is thus of fundamental importance because everything else is dependent on it. To put it rather strikingly, in philosophical enquiries, a philosophy of mind has priority over truth-theory. Further, and that is important, this theory of mind has to strike a balance. Firstly, it must be a theory of mind about which we should feel that it could be plausible despite the modification it might demand in our commonsensical or the intuitive view of mind, and secondly it must be powerful enough to give us the required analytical tool, namely a method to move from the preferred to the problematic sentences. From this it should be evident that in the Critic of Pure Reason, Kant was after the sort of a theory of mind which we are talking about. Whether he succeeded in his attempt is not the issue; what we are interested in for the present is the very nature of his program. Further, Kant did not intend that theory of mind to be empirical; fortunately for philosophy, the Critic of Pure Reason did not turn out to be treatise in experimental psychology.

In the foregoing I have not read into Kant; that it is there in his writings can be elicited by considering the nature of the laws concerning human mind which Kant formulates. One of the laws of his theory is that the Euclidean three dimensional space had to be the space in which human mind localizes, particularizes, and differentiates individual appearances. A crude but fair clarification of this would be that human mind was bound to impose this three dimensional Euclidean space as form of intuition and that it is bound to do so because it is constrained or constructed by itself to objectify appearances. To speak the Wittgensteinian lingo, the inexorability is of mind and not of the world; see (12.1.118). this seems to be reason for Kant's not holding that real or objective space will have to be three dimensional Euclidean. His point is that we do not know the structure of real space a priori. It might be that it is multi-dimensional non-Euclidean. We will have to concede that, while admitting that our visual space is only what it is, namely three dimensional Euclidean. This is the way things are with us, or this is what we are.

[10] Both these arguments are equally thrilling; but there are better scholars of either of these great minds who are excited about one but not the other.

[11] Which when freely translated would read as: "Generally it is the nature of every advance that it appears much greater than it really is."

[12] This somehow is missed by those who subscribed to the Remarks view of mathematics as well as those who following Russell thought that the view is the "surprisingly insignificant product of a sparkling mind", to quote Kriesel. Even if Russell is right, it does not affect the stature of Wittgenstein because only mediocres excel continuously, and a genius excels only once.

[13] Wittgenstein remarks in (12):

> "Arithmetic is the natural history of mineralogy of numbers; but who talks like this about it? Our whole thinking is permeated with this idea,"

and

> "it is the remark of mathematical alchemy that mathematical statements are regarded as statements about mathematical objects, and so mathematics as the exploration of these objects."

[14] i.e. life as it is lived daily.

[15] Pace Tractatus according to which there are no negative facts; nor are there any negative elementary sentences. Throughout his philosophical career Wittgenstein held that nature, after all, was positive.

[16] I use 'representation', 'picturing', and 'depiction' synonymously; keeping them in quotes may be in order, but I shall not do so as I assume that by now it has become a commonplace that they should be used with single quotes. Wittgenstein's idea is a technical one, and is informed by the mathematical practice of finding a model of one system (structure) in another, for example non-Euclidean geometry in Euclidean geometry, and geometry in the arithmetic of real numbers. This is called "abbildung" of one system in another; see for instance 4.015 of Tractatus where Wittgenstein talks of "logik der Abbildung". He is also influenced by Hertz, cf. Principles of Mechanics, who takes the entire physical theory as a "bild" designed to provide a "darstellung" (representation) of the physical phenomenon. Here "darstellung" is used as a synonym for "abbildung". Wittgenstein's "darstellung" is to be contrasted with Kant's "vorstellung"; the latter has psychological affiliations in virtue of being located in an individual mind. The former is public representation where both the relata are public, and stands in contrast with "vorstellung" where one of the relata is private. Thus what we mean by 'our representation' is 'the representation of our language'.

[17] Whether the meaning of the personal pronoun in plural is the same in all the occurrences in this sentence is really problematic. This sentence is clearly anomalous; and this itself can be used in rejecting the hypothetical argument, or in brushing aside the thought-experiment.

[18] It may be noted that for Wittgenstein that the tennis ball is round is a statement, and that the Earth is round is a hypothesis.

[19] This idea persists in Wittgenstein's thought right from the beginning; see Tractatus 6.342:

> "... the fact that it (i.e. the world) can be described by Newtonian mechanics asserts nothing about the world, but this asserts something, namely that it can be described in that particular way in which as a matter of fact it is described."

This is not a truth about the world; it is a truth about man. It is not a fact of the natural history of the world, but of the natural history of man.

[20] See Endnote 9 above. They are perfectly intelligible, but are of no use. Though a Pagan I know what it means to live a Christian way of life; only I cannot live it. Practice does not limit understanding.

[21] Incidentally, double verb sentences like the first in this example are deceptive; Wittgenstein offered his own analysis of such sentences.

[22] Even this may be an arrogant claim; all that we are inclined to subscribe to is that after all those brilliant results are arrived at we have reached a point, a vantage point perhaps, from where we can glimpse through the nature of a set.

[23] Philosophers should be interested only in the semantic aspect of the issue; concepts may have other affiliations which would interest say psychologists.

[24] The normative character of rules thus makes man a normative being with a natural base - as in Aristotle.

[25] This is a bit Spinoza-like. Did not Spinoza maintain that what is meant by that man is a value-oriented being is not that there are values and man pursues them, but that what man pursues are values!

[26] (12).V.18.

[27] What belongs to the natural history of man is not mathematical sentences; they are not descriptive of the human way of acting, but the fact that we use sentences as we do, this is that we do shape our concepts thus, and that we behave thus.

[28] See his Metaphysics, Huchinson, 1965.

[29] The following passage from Wittgenstein's (14) of which the ethno-methodologists are very fond of might be invoked to cast doubt on my interpretation; so in my defense I may explain the passage:

> "The very idea of wanting to explain a practice ... seems wrong to me. All that Frazer does is to make them plausible to people who think as he does. It is remarkable that in the final analysis all those practices are presented as, so to speak, pieces of stupidity."

As I see it, what this remark amounts to is this: first, explaining a practice is wrong not because the very existence of a practice analytically shows that it is rational, in which case the explanation would not be wrong but simply useless, but because the rationality or the otherwiseness of a practice is independent of its existence; this is so for the simple reason that irrational practices, then, do not exist. Explanations go wrong, as Wittgenstein is saying that they have gone wrong in the case of Frazer, because of the prejudice of understanding, because of the constraints which one's constitution imposes on him. As Wittgenstein wondered:

> "what a narrow spiritual life on Frazer's part! As a result, how impossible it was for him to conceive of a life different from that of England of his time?"

It may be added that more often such explanations commit what may be called the Socratic Fallacy which can be schematically presented as: I have a vague (intriguing) picture of an object, therefore I have a picture of a vague (intriguing) object.

[30] There are far too many ontological problems involved in the linkage between contingent constitutivity on the one hand and persistence and change on the other; but none of them should bother us here.

[31] Pace Tractatus where Wittgenstein stripped off the world from all essential connections or necessary relations and held that objects form the substance of the world.

[32] This may be Wittgenstein's way of unpacking Goethe's equally famous or notorious line from the Faust: "… in the beginning was the deed." Wittgenstein echoes this Goethean idea in (15.46) and (3.1.546).

[33] Wittgenstein thought that he was "dealing … with the Kantian solution to the problem of philosophy," and not, it may be noted, with a solution to the Kantian problem of philosophy.

[34] I am assuming that it is the visual system and not the physical system which is at issue here. This is the place where we should take note of Hintikka's brilliant analysis of the concept of colour in the Tractatus. In their Investigating Wittgenstein the Hintikkas interpret the Tractatus to be suggesting that the colour concept is not represented in language (either in our language or in the "language" of the butterflies) by the class of colour predicates. On the other hand it is represented by a function C which maps points in visual space onto colour space, this is to say that what we call the visual system onto the physical system. Then the logical form of sentences "This patch is red" and "That patch is green" - in the language of the butterflies "This patch is this black" and "That patch is that white" - would be $C(a)=r$ and $C(b)=g$, where r and g are two separate "objects" red and green. From this it follows that something's being both red and green is logically impossible because a function cannot have two different values for the same argument. This explains the problem of colour incompatibility in our case as well as in the case of the butterflies; obviously the same object cannot have two different shades of black or two different shades of white. And the physical system remains the same for all of us including the butterflies.

[35] These and their ilk constitute a form of life. A form of life is almost a biological concept. There is one form of life which characterizes our species and reflects our species. There are other forms of life which characterize other biological species.

[36] At this point Wittgenstein's thought can be hooked to the Frankfurt school, specially to Habermas; no such attempt seems to have been made till now.

[37] I do not know what different interest Brouwer had, but some of his prejudices are not to be shared by us because they are more personal rather than intellectual.

[38] Without exception 'philosophy of …' invariably remained irrelevant to each of those disciplines which can go to fill the hyphen. That is why that expression needs to be taken into cognizance only when the blank is filled by the name of a person. 'Philosophy of Plato', 'philosophy of Hegel' are perfectly intelligible expressions. But it is difficult to figure out what 'philosophy of the social sciences' or 'philosophy of religion' means.

[39] Wittgenstein on Foundations of Mathematics, Duckworth, 1983. What I am saying here should be taken as a supplement to, and not as a substitute for, what he says there.

[40] Mathematics is an (the) exact science only in its method. There are no exact sciences, but only exact methods.

[41] As it is evident, here I used Wittgenstein's thoughts ignoring their chronological order. It is immaterial, for the purpose at hand, when he thought a particular thought. What is important is whether his thoughts form a coherent whole or not. Moreover, I am not a "Trinitarian" to isolate the early, the middle, and the later Wittgensteins. I take his thought as a "Unitarian' would take it, that is relegating history into the background. For him, philosophical enquiry is not an historical investigation. Further, I am interested in his

contribution to philosophy, and not in the evolution of his thought. In defense of this I might cite a passage from his (14):

> "The historical explanation, the explanation of an hypothesis of development, is only one way of assembling data ... it is just possible to see the data in their relation to one another and embrace them in a general picture without putting it in the form of an hypothesis about temporal development."

But for this Wittgenstein may not have published the Tractatus; he at most may have consented to publish what we now have as Note Books. It is commonplace that Note Books contain his thoughts in the chronological order, and the Tractatus contains them in their logical order. I am not saying that this is all the difference between them; obviously, there are ideas which figure in one, and which Wittgenstein rejected by the time he came to the other, for example that logical constants are names. It seems that this rejection itself is to be understood in terms of the order imposed on those thoughts in the Tractatus. In any case, if it is so wanted, this may be considered as a contribution towards a Wittgensteinian philosophy rather than Wittgenstein's philosophy.

Chapter Three : On Generality

Though exercising the prerogative of a satirist, Swift caricatured Hobbes who, as the professor in the School of Languages at the Academy of Lagadoo, sought to reform language by abolishing all words in favour of minuscule things, and also laughed at the natural historians of language of the reconnaissance period, no one - not even Hobbes, because he too at a certain point of his intellectual life was fascinated, if not bewitched, by Euclid - ever doubted that it was in the nature of human thinking itself that its natural thrust was invariably towards generality. The natural history of man (in the ordinary sense of the term) shows, contra the Laputan academics, how with the invention of language man liberated himself from the constraints of things, and how with the advent of mathematics from the constraints of language itself! The belief of the twentieth century man is that nothing can slip from his mathematical net, indeed that anything that cannot be caught with it does not exist. This belief finds expression in the Tractatus at 5.526.

> "We can describe the world completely by means of fully generalized sentences, that is, without first correlating any name with a particular object. Then, in order to arrive at the customary mode of expression we simply need to add, after an expression like 'There is one and only one x such that …' the words 'and that x is a'."

I shall devote this chapter first to understand this claim and then to evaluate it in order to judge its legitimacy and to fix its limits. Because several hard perennials of philosophy are involved in this claim, and as all of them may not be traceable completely until the whole ground is dug up and upturned, I may have to carry on this task with restricted aims. I shall not thus be able to explore any issue, however important it otherwise might be, which has no immediate relevance to our task here. And in carrying out even the restricted task which I set myself to, I shall assume a minimal acquaintance with the standard quantification theory and some meta-theoretic results.

In the recent logico-philosophical investigations the two issues,

(a) what is (the nature of) a general(ized) sentence?

and

(b) what is the general form of a sentence?

are separated; at least, they are separate issues. For example Strawson in his editorial introduction to Philosophical Logic[1] creates the impression that this discipline is exclusively concerned with (b). one of the central theses of the Tractatus is that each sentence, whatever be its surface structure, in its deep structure it is a general sentence. The "essence", that is the logical structure, of a sentence is its generalized form (5.471); and this mechanism is geared to provide rubrics for stating: this is how objects stand (94.51).

But what is an object! We know very little about objects in the sense that the ontic category to which they belong eludes us. That is why, as Wittgenstein remarked to Waismann[2]:

> "(I)t makes no sense to ask whether objects are thing-like, whether they are something like property, or are relations and so on."

Nevertheless, it is a part of the little which we know of objects that they exist. This is the same as to say that their existence is assumed in our linguistic activity.

> "What looks as if it had to exist is a part of our language. It is a paradigm in our language game, something with which comparison is made. And this may be an important observation; but it is nonetheless an observation concerning our language game - our method of representation." (3.1.50)

To credit it to one to whom it is due, this in fact was a discovery of Aristotle, who first tried to show how, though there can be a false statement about the existent, there cannot be a true statement about the non-existent, and then made it one of the foundations stones of his syllogistic[3]. And it finds embedded in the standard quantification theory, which as is well known takes its models from only domains which are not empty. There are scores of suggestions as to how the standard quantification theory can be redeemed from this assumption, which in literature is referred to as the "existence assumption"[4]. I shall not, however, discuss any of those suggestions unless and until it cuts across the point under discussion here. The point with which I am concerned here, to state it succinctly, is that the rubrics of the standard quantification theory capture only two of the three mechanisms required for a successful referential use of language whose structure coincides with it. These three mechanisms are:

(i) the existence of the referent

(ii) the constancy of reference, and

(iii) the uniqueness in referring.

The standard quantification theory does not provide for the last of the three.

Having noticed this failing of this theory, Russell tried to overcome the lacuna by augmenting it with theories of identity and definite descriptions; as it is well-known, Wittgenstein had serious objections to Russell's theory of identity[5]. The essence of what Wittgenstein had to say about identity in the <u>Tractatus</u> was summed up in 5.53, 5.5303, and 5.5352; to cite them together,

> "Identity of objects I express by identity of signs and not by using a sign for identity. Difference of objects I express by difference of signs. Roughly speaking, to say of two things that they are identical is nonsense, and to say of one thing that it is identical to itself is to say nothing. The identity sign, therefore, is not an essential constituent of the concept-script ... (P)eople have wanted to express 'There are no things' by writing '~(Ex)(x = x)'. But even if this were a sentence, would it not be equally true if in fact 'there are things', but they were not identical with themselves."[6]

Wittgenstein seems to have noticed that what is embedded in the passage cited above, though expressible in the standard quantification theory with a non-standard interpretation, cannot be expressed in that theory with the standard interpretation; this is the same as to say that the Frege-Russell way of reading the quantifiers does not fit with his view of identity. And here-in lies whatever is central to the issue. Because one's understanding of the concept of identity and his interpretation of the quantifiers and variables go hand in hand[7], given the Frege-Russell way of reading the quantifiers and the variables - i.e. on what has come to be called the inclusive interpretation - and also the Frege-Russell way of extending the standard quantification theory to include identity, the following schemata come out valid.

(i) (x) Fx → (y) Fy, and (ii) (Ex) Fx → (Ey) Fy.

Because of this, this is to say that because in this interpretation all the variables are interpreted to be ranging over the entire domain of interpretation, numerical sentences require identity. For instance, that there is only one individual with a certain property P can be represented as (Ex)(y){Px. (Py → x = y)}, and that there are two individuals with the same property as (Ex)(Ey)(z)[Px.Py.{Pz → (x = z∨y = z)}]. On the other hand if we interpret the variables assigning different domains to different variables[8] - or in a weaker fashion allowing different elements of the domain of interpretation to be the values of variables while allowing all the variables range over the same domain - the very same can be represented respectively as (Ex)(Px) and (Ex)(Ey)(Px.Py).

Yet, quantification theory even so interpreted but not extended in some way or the other fails to incorporate the mechanism required to capture uniqueness. I may explain this using the current idiom without attributing either the idiom or the explanation to Wittgenstein. Consider for example two models M and M' with the same domain D, and let D be {e,e'}. Let it be that the monadic predicate P in M defined over the subset {e} of D, and in M' over the subset {e'}. Now (Ex)(Px) comes to take as its models both M and M'. To paraphrase a well-known result about the quantification theory, it takes every M+ which is isomorphic with M as its model if it takes M as its model. To illustrate the same for the benefit of those who are not iniated in this spurious ornamentation, let us assume that there is nothing else in this universe except your bedside bookshelf and my bedside bookshelf. Let it also be the case that each of these shelves contains two books say the Tractatus and the Investigations; and let the cover of my copy of the Tractatus be red and the cover of your copy of it to be green. Let the covers of your copy of the Investigations as well as my copy of it both be red[9]. Now, that there is a red book - a book with red covers, i.e. - on the shelf, which when telescopically symbolized in the language of quantification would read as (Ex)(Rx.Bx); and this is true both about your shelf as well as my shelf. It is true of uniquely every shelf which has at least one red book on it, and is also true of any shelf which has at most one red book on it. But it is not uniquely true of any one shelf among the two shelves. The Frege-Russell referential interpretation of the variables is geared to make it a truth about all those shelves which have at least one red book on them; and Wittgenstein's referential exclusive interpretation is attuned to make it a truth about all those shelves which have at most one book which is red. His trust was in that, because our referential use of language assumes the existence of the referents, Frege-Russell representation was needless. His belief was that existence of at least one red book assumes that the interpretation of the variables should be carried out with the intention of guaranteeing the existence of at most one red book. As his interpretation assumes the existence of at least one red book, it follows that under his interpretation, asserting the sentence in question amounts to an assertion to the effect that there is at least one and at most one red book on the shelf. Yet that would be true of both your shelf as well as mine; and in order to make it not only *true of uniquely* about both the shelves, but *uniquely true of* one of them, it might be suggested that we can bring into action additional variables and predicates, and reformulate the same as, for example

(Ex)(Ey){(R'y.Sx).(Rx.Bx).Sxy};

this is to say that by showing that the sentence

There is a red book on the shelf

is only an elliptical version of the sentence

> There is something such that it is shelf and is of Rao, there is something (else) which is a book and has red covers, and this something (else) is stacked on a shelf which belongs to Rao.

The puzzlement one faces here, to formulate it with some focus, is that whether the term 'else' should figure in this sentence which, as it can be noticed easily, is intended to bridge the gulf between the two senses of being true. To disclose the outcome of the following discussion, the issue is whether the other sense of being about is brought in by variables or else by predicates. Wittgenstein insists that it is due to the efficacy of the variables, bound variables to be specific; and Frege and Russell emphasize that it is the work of the predicates. Thus assuming the Quinean precept that ontology is determined by bound variables, what needs to be settled is whether being uniquely about is an ontological matter or else is an ideological affair, for ideology can analogously be understood as something determined by predicates[10].

Russell, while sticking to his interpretation of the quantification theory, wanted to secure for it what it would come to have under the Wittgensteinian interpretation by introducing into it the theory of identity. He tried to show that in the extended quantification theory we would be able to state that at least one and at most one individual has a certain property. And to capture uniqueness too Russell brought in the theory of descriptions, thus further extending the quantification theory. It was the claim of Russell that in this fully blown theory of quantification all the three mechanisms required for a successful referential use of language, namely existence, constancy, and uniqueness can be captured. But the fully blown theory has not only these desirable features, but has in addition what Wittgenstein considered to be a "logical trick" built into it. He thought that these features were undesirable on the count of their incorporating an illegitimate concept of identity. Reserving a detailed consideration of identity to the next chapter, I may note here that the theory of descriptions which was based on that concept, Wittgenstein seems to have thought, goes against the very foundations of his semantic system - to be specific, his belief that the class of referring expressions and the class of expressions which have sense are mutually exclusive[11]. In view of the equivalence of $(Ix)Ax$, and[12] $(y)(Ex)\{Ax.(Ay \rightarrow x = y)\}$, Wittgenstein seems to have thought that Russell is blurring the distinction between terms and sentences[13].

At this stage we may place ourselves in the second decade of this century and be concerned with what, after all, the outcome would be if we decline to extend the quantification theory the way Russell does and be satisfied with reinterpreting it on the lines suggested by Wittgenstein[14]. Obviously, this is what Wittgenstein was talking about in 3.24 of the *Tractatus*.

> "In fact a generality sign contains a prototype, it leaves something indeterminate, and descriptions in terms of generality are devoid of uniqueness."

(*Descriptions*, here, does not mean definite descriptions, but *representations*.) Working on his insights we may note that the absence of the rubrics for capturing uniqueness in the standard quantification theory is not due to any inherent defect of the theory (with or without the Wittgensteinian interpretation of the variables.) Uniqueness does not belong to the realm of syntax or even to the realm of semantics; it falls into the area of pragmatics[15]. I am not suggesting that Wittgenstein invented the distinction between these three areas of enquiry concerning language; all that I entertain is that he seems to hold views which amount to tacitly projecting such a distinction. A sentence could uniquely be about a fact, and not about any fact isomorphic with it, only in the context of its use[16]; this insight, of course, can be credited to Wittgenstein. It is already present in the Tractatus, though he devoted much of his subsequent writing to it. Thus this point demands some attention, and a clarification.

To meet that demand we need to harp a little more on the distinction between uniquely being about and being uniquely about; for convenience let us refer to the two senses of being about respectively as: uniqueness-1, and uniqueness-2. It is in the sense of uniqueness-1 that we say that physics, for instance, is about the physical world[17]. In 6.6431 of the Tractatus Wittgenstein is remarking that:

> "The laws of physics, with all their logical apparatus, still speak, however indirectly, about the objects of the world. We ought not to forget that any description of the world by means of mechanics will be of completely general kind. For example, it will never mention particular point-masses. It will talk about any point-masses whatever."[18]

Because it is geared, or intended, or used to capture only uniqueness-1, a general sentence cannot be treated as truth-functional compound[19]. Though Wittgenstein was repeatedly and explicitly asserting that he does not treat general sentences as truth-functional compounds, if it has come to be believed by a large number of his readers in the English speaking world (at least) that he does treat them so, it is - as it seems - due to a misunderstanding which Russell perpetuated through his introduction to the Tractatus. In fact, as Wittgenstein did point out there, it was Russell who held that position, and not he himself.

In the first chapter I pointed out how Russell is habituated to attack a philosophical position which he himself held at an earlier period while imputing it to others. He does it once again in his introduction to the Tractatus by imputing to Wittgenstein the view of general sentences which Wittgenstein in his turn is crediting to Frege and Russell themselves. (There is more to this than what I am pointing at here, and I paid greater attention to it in my (9).) This can induce one to wonder whether Russell has gone through the text with the care it deserved before writing the introduction; it could be that he wrote it relying more on his talks with Wittgenstein. Anyhow at 5.521 of the Tractatus, Wittgenstein is saying:

> "I dissociate the concept all from truth-functions. Frege and Russell introduced generality in association with logical product and logical sum. This makes it difficult to understand the sentence '(Ex)Fx' and '(x)Fx' in which both these ideas are embedded."[21]

Thesis (satz) 5 and the comments on it in the Tractatus show Wittgenstein's serious engagement with the problem of general sentences, the outcome of which is his solution to it through the N-operator. He devotes thesis 5 and the comments on it to consider the form of general sentences (and 6 and comments on it to understand the general form of sentences as such) using the N-operator which is a generalization of Sheffer's Stroke. Thus $N(\bar{Z})$ is the result of applying the operator of joint denial to $\bar{Z}$, where Z is a set of sentences; the bar is intended to indicate that $\bar{Z}$ represents all of its values, as 5.501 specifies. $N(\bar{Z})$ then comes to be true when all the numbers of Z are false. Then

> 5.02 If Z has as its values all the values of a function Fx for all values of x, then $N(\bar{Z}) = \sim$ (Ex)Fx.

Given that negation here is normal, it is evident that $N\{N(\bar{Z})\} \equiv$ (Ex)Fx, provided $\bar{Z}$ has, as its members, all the substitution instances of Fx. As the N-operator does not in any way affect the interdefinability of the quantifiers, it may create the impression that it solves the problem at hand. But it really brings us to square one.

On this reading of the quantifiers, to be able to express say (Ex)Fx one needs to specify, as the members of $\bar{Z}$ all the values of the function Fx; and this can be done either by enumerating those values if they are finite in number, or by writing (Ex)Fx. The purpose of the quantifier (Ex) thus is to indicate the function that is being handled. If this is indeed what Wittgenstein had on his mind, he seems to be suggesting that Russell's understanding of the quantifier (Ex) confuses the indication of the values (variantly put in Russell's terminology, the propositional function) and the joint denial of those values (to put it differently, again in Russell's terminology, the logical sum of the negations of those values.) That apart, this device falls short of providing rubrics for expressing all the generalized sentences; it does not cover multiply quantified sentences like (Ex)(Ey)(Fxy). When Faa, Fab, ..., Fba, ... are the values of Fxy, N(Fxy) will have the same truth-value as ~ (Ex)(Ey)(Fxy), and N{N(Fxy)} will be materially equivalent to (Ex)(Ey)(Fxy). Now reiterated applications of the N-operator on these will yield only equivalents to either of these. Thus the operator stands in need of a modification so that it may clearly indicate the sentential function in question, this is to say that, indicate the sentential function which is the argument of a specific instance of the operator. For instance, if Nx(Fxy) were to indicate the joint denial of all values of Fxy, for only all values of x, (Ex)(y)(Fxy) can be expressed as N(Nx[Ny{N(Fxy)}]), which so to say, is the N-counterpart of ~(x)~(y)~~(Fxy). In short, as 5.501 suggests

$$(Ex)Fx \equiv N\{Nx(Fx)\} \equiv N\{N(Fa,Fb,Fc,\ldots)\}.$$

and

$$(x)Fx \equiv Nx\{N(Fx)\} \equiv N\{N(Fa),N(Fb),N(Fc),\ldots\}.$$

The original N-operator and its modified version in Nx are oriented towards disparate purposes, which, as Wittgenstein seems to have thought, are presumably fused or confused by Russell. In fact, where N is an operator - of joint denial, obviously - Nx behaves as if it were a function whose arguments are quantified formulas or generalized sentences. Several characteristics of the sign of generality mentioned in the Tractatus may gain transparency from this, for instance how it figures as an argument (5.523), and how it "indicates a logical prototype" as Wittgenstein remarks in 5.522.

Russell's misunderstanding may even have been occasioned by the pithy remarks of Wittgenstein concerning the relationship between generalized sentences and the elementary sentences, for instance 4.411 of the Tractatus,

> "Understanding of general sentences palpably depends upon the understanding of elementary sentences."

Identifying elementary sentences with sentences which traditionally are referred to as singular sentences - which, without being unfair to him, cannot be attributed to Wittgenstein - Russell might have come to believe in what he did believe. However, Russell's belief is unwarranted because what Wittgenstein is hinting at in that remark is not that general sentences are either sums or products, but that understanding elementary sentences provides a clue or key to understand generalized sentences because their structural components are the same - or, to put it differently, because they are all made up of components of the same logical categories.

> 5.454 In Logic there is no distinction between the general and the specific.

As the expressive power of a sentence is derivative of its structure, the general and the specific are non-distinct with respect to their expressive power; this is to say that the generalized and the elementary are not different in their capacity to represent, a capacity which they come to have solely by virtue of their logical constitution, this is to say that due their being constituted by logical components. Being constituted by logical components, namely predicates, variables, and quantifiers, gives sentences - be they general or elementary - their logical form in which the representational powers remain constant and everything else is varying. This is a bit intriguing because apparently quantifiers do no deem to go into the composition of the elementary sentences. Thus the clarification of Wittgenstein in 3.312 of the Tractatus concerning the invariant element which logic secures to a sentence; and this needs to be understood in the light of 5.547,

> "An elementary sentence really contains all logical operators in itself. For 'fa' says the same thing[22] as '(Ex)(Fx . x = a)'. Wherever there is compositness argument and function are present, we already have all the logical constants."

Wittgenstein's view is that if we proceed on the lines suggested by Russell, showing how the definite descriptive phrases do not augment the expressive power of the extended quantification theory, we would arrive at a point where we would realize that names too do not augment the expressive power of our language in virtue of the coincidence of its logical structure with the standard quantification theory with exclusive referential interpretation of the variables as suggested by Wittgenstein. This should have been known to Russell more than anyone else; and Wittgenstein noticed it much before Quine. Names indeed are the sort of variables which we use in existential instantiation; they are what Quine calls the flagged variables. It may be noted that in

(Ex)Fx

Fa

we are using a as a variable[23], just like y in

(Ex)Fx

Fy

All this has a consequence which may surprise some and shock others. It is that there are no genuinely singular terms or singular sentences in the sense in which these are usually understood; singular sentences in the strict sense of the term may be possible only in the realm of pragmatics - in the actual linguistic practice in which sentences are contextualized to the point of their becoming unique in the intended sense. This should naturally be the initial reaction. But we are, after our enquiry in the last chapter, so to say a little wiser, and hence may begin seeing things anew. We may now see how to account for the way in which we have been using singular terms and singular sentences in general, and names in particular. We are now in a position to work out a natural history of naming.

What emerges till now is that the logical behaviour of names is precisely the same as that of the variables of instantation; and we have noted that this view goes well with the idea that our descriptions and representations are structural only. A sentence is descriptive only in virtue of its being structurally isomorphic with the corresponding fact; as such the descriptum is the corresponding fact and not any of the objects which configure in it. The general form of a descriptive sentence - namely 'such-and-such is such-and-such' - links the sentence and its constituents to the components of the corresponding fact, which are objects, only through the shared form. This is the way in which a sentence comes to be about objects; it is about them only indirectly. Now, it can be noticed that if 'the such-and-such is such-and-such' is true at all, it is true of 'the such-and-such like is such-and-such like'[24]. Does it not then follow that there is no particularity involved in the objects of reference either. Objects are values of bound variables, after all. Even if they cannot be described[25], they can at least be named, which is what Wittgenstein is saying in 3.221. Cannot we, then, at least pickup objects of the world by means of names? If we share the Wittgensteinian style of thinking, which we should if we are to understand him as he wants us to understand, our reply should first be: let us note what is being sought after; what is it that is being asked, that is. Let me attend to this.

A name has no unique semantic relationship with the object in its particularity, individuality or whatever. The relationship itself is epistemic and pragmatic; it might even be the case that the nature of this relationship is determined by specific interests in isolating and naming disparate chunks of the spatio-temporal mass given to us, as it were. That apart, the relationship in question does not seem to be unique for a trivial as well as a profound reason. The trivial one may be illustrated by an example. 'Socrates', for instance, does not only refer to the biped who bullied intelligent young men in the Athenian market place, but also the biped whom Xanthippe judiciously treated, ..., the biped whom Plato respected most but failed to turn up when he voluntarily drank hemlock, ..., the biped whom every philosopher admires without ever following him, ..., ad infinitum; the dots are not of laziness, and the last term was meant with all seriousness[26]. As unity and particularity go together, it is incomprehensible, at least as long as we confine our enquiry to logic and semantics, how to unify all these apparently disparate bipeds to individuate a particular biped and hook him by the name 'Socrates'.

If so, what would happen to 3.203 of the Tractatus? Have I, in the first chapter, expended my labour in vain? At one point Wittgenstein creates the impression that had this question been posed to him he would have been inclined to give a positive reply. In his 1932 conversation with Waismann[27], Wittgenstein is reported to have admitted that in the Tractatus period he thought that ostensive definitions confront names and the world (objects?), this is to say that in a certain context, the speech act using the sentence say 'This is Mother Teresa' marries off the world and language monogamously. But Waismann says that Wittgenstein went on to clarify:

> "when I wrote Tractatus, I was unclear about logical analysis and ostensive definition. I then thought that there is a linking up of language with reality."

May be, as nobody should be granted a monopoly over a correct understanding of his own earlier thoughts and actions, we can take these remarks of Wittgenstein in a different light and doubt that they amount to a rejection of

3.203, which is the way in which they are usually understood. In doing so we will not be violating any Wittgensteinian norm of philosophical interpretation. Did not Wittgenstein entertain the possibility of one's understanding another's thought better than he himself understands it! One has only to skip through his (15) to not how he does; and I do not think that Wittgenstein ever had exclusive access to this gift. In any case, it seems that in the conversation under consideration neither Wittgenstein nor Waismann saw through clearly the implications of the remark quoted above. It appears that there are, in so far as I can isolate for the present, two distinctly different issues which are being jumbled up by Wittgenstein. The first of these is: how are names related to objects? And the second is: what kind of a category of objects do names refer to[29]? The first is a semantic problem, and the second is an ontological issue; also the second is an issue in substantive ontology and not even formal ontology which is one of the concerns of Wittgenstein in the Tractatus. 3.203 is not a remark in ontology, least of all substantive ontology. It is a statement of semantics. Further, it is important to note that my characterization of 3.203 as a semantical statement does not go against what I said earlier, namely that there may not be any feature of the world which is not revealed by language. Because descriptions are only structural, none of the structural features of the world will go unrepresented in language. If individual unity and particularity of objects of reference are not revealed by the syntactical or the semantic features of language it is because it is something which belongs to the realm of pragmatics, this is to say that it belongs to the applications and uses of language. Of course, that we cannot use language to represent the world in its totality and unity is obvious; we are so to say condemned to use language to represent the world sub species. We are logically constrained to be content with partial descriptions or descriptions of partial worlds!

Each of our partial theories has its own way of individuating or particularizing, or to put it differently its own method of constructing a class of identical unit classes[30]. Occasionally, it may, as it does, turn out that several partial theories have the same way of individuating particulars incorporated into them. This happens when the language in those theories gets applied or used in the same fashion. It could even be that in some theories objects come individuated as Goddess Athena did from the forehead of Zeus fully grown and fully attired. Fundamental particle theory is one such, perhaps. In such cases the theory, providing a description of that part of the world which those objects make up, will not be incorporating a principle of individuation because no such is needed. Theories receiving model theoretic semantics do not, as they need not, care for principles of individuation; they assume individuated objects as referents of their referring terms. There is nothing wrong in this, because what is given to us prior to our linguistic or theoretic intervention is an undifferentiated and non-descript mass revealed to us through our sensory apparatus[31]. Carving out particular individual chunks out of this mass is a linguistic affair (as the Boston Buddhist, namely Quine would say), falling under the purview of pragmatics (as the Indian Buddhist say).

This has two important messages. First, the principles of individuation do not come apiece; this is to say that each object is not individuated separately. Perhaps this is why a sentence in which a name is linked by the copula to a definite description is still informative whether it is true or false. The copula in such a sentence cannot always be taken as identity[32]. Secondly, the principle of individuation embedded in a theory - irrespective of the fact whether it receives an explicit formulation, or else remains tacit while being used continuously in the interpretations of that theory over its intended models - individuates all and only members of the domain of interpretation of that theory[33].

If a name refers uniquely (in the sense of uniqueness-2) to an object in the domain of interpretation, this is due to - as we have come to notice - its being able to figure in the place of a bound variable of a theory in which that name occurs as a primitive term together with the principle of individuation incorporated in that theory. Thus Tractatus 3.328 which says:

> "If a name had reference on its own, it is difficult to see why it fails to refer."

This means that autonomy of nomial reference does not sound plausible. In the beginning there was not the word, and there is a need to rewrite that part of the Genesis[34].

As if moving in that direction, Wittgenstein notes in his (10) that

> "(t)he referent of a name is not the thing we point to when we give an ostensive definition of the name."

On the other hand, it is the object which is individuated by the principle of individuation used in the context of the ostensive definition. This means that the referent of a name is individuated in a given theoretical or practical context, in the broad senses of the terms[35]. As such what a theory refers to is theory-dependent[36]. Further, because of these considerations whatever a name refers to, it does not refer to in isolation from all the other referring expressions of the theory. We cannot fish the objects of the world by a nomial rod; we have to net them in a theory. Otherwise, objects and not facts would be the discernibilia of the world; but this goes against the Wittgensteinian creation-myth in the very first sentence of the Tractatus, namely that the world is the totality of facts (and not of objects, as has been believed till now). One may, of course, start picking up Quinean capsules here. But I do not propose to do so; I mentioned them only to indicate that they have ancestral Austrian affiliations of which Quine is not aware, it seems. So I will move to consider, in brief though, a couple of other ideas relevant to the issue at hand.

What I have been saying about names does apply to predicates too, I trust. As I intend to consider this point in the next chapter also, here let me just note that there is no reason why we should not consider properties which are referents of predicates, though all predicates may not have referents, as spatio-temporally, or spatially, or temporally scattered objects. There seem to be sentences which we use, and use not infrequently, creating the impression that we, in our daily life, do in fact tacitly subscribe to such a way of carving out the world. 'This family has ruined the country', 'The current inflation will swallow all of us', for instance. And, after all, a broken jug - as the New England philosophers would say - is still a jug though a bit of it is here, and another there, and some other bit not traceable at all; and I do, as I believe legitimately, talk of my career even though it is only a set of scattered occasions with a dozen of intermissions and in-betweens. Spatial contiguity or temporal continuity should not go as essential ingredients of individuation, which anyhow is a matter of semantics or pragmatics[37]. This is indicative of how individuation of properties is theory-relative or is dependent on the context, and thereby belongs to the realm of practice[38]. This similarity between names and predicates, as it seems, induced Wittgenstein to believe that the general form of a sentence can be captured only by a totally generalized form, this is to say that by treating predicates as instantiated predicate variables (of a language whose logical structure coincides with that of second order quantification theory with an exclusive interpretation of predicate as well as individual variables.)

> 5.5261 A fully generalized sentence like every other sentence is composite. This is shown by the fact that in '(Ex,ϕ)(ϕx)' we have to mention 'ϕ' and 'x' separately. <u>They both, independently, stand in signifying relations to the world</u>, just as is the case of ungeneralized sentences[39].

- like 'Socrates is taller than Simmias', or even 'Socrates is snub-nosed', perhaps. (It may be noted that Wittgenstein would not have permitted sentential quantifiers, or quantification over sentences resulting in expressions like (P)(P → P), where P is a sentential variable, because the substituends for the variable will have to be names. Such a generalization would force us to treat sentences as names, this is to say that as expressions with referents; and this goes against the very foundations of Wittgenstein's views on semantics.)

If the respective behaviours of the argument-part and the function-part of a sentence are so much alike, why after all they are presented in a sentence by different kinds of signs! Why not instead of transcribing as (Ex,ϕ)(ϕx), transcribe the same as (Ex)(x,y), or else as (Eϕ,ψ)(ϕ,ψ)! Let me try to draw out at least a Wittgensteinian, if not Wittgenstein's, answer from his ideas scattered all over the Tractatus in the form of cryptic remarks full of insight. That we need to use two different kinds of signs exhibiting the sentential form - one for the argument part and another for the function part, to use the Fregean terminology - follows from Wittgenstein's views on (a) identity, (b) mode of signification, and (c) mode of sign-production. Considering (c) first, generalizing on 3.322 of the Tractatus we may note that:

> "... our use of the same (kind of) sign(s) to signify two different (kinds of) objects can never indicate a common characteristic of the two, if we use (them) with two different modes of signification. For the sign(s), of course, (are) arbitrary. So we could choose two different (kinds of) signs instead ..."

In the two immediate comments on this remark Wittgenstein draws out our attention to how we, in our common parlance, frequently err in sign-production by using the same kind of signs with different modes of signification, and different kinds of signs with the same mode of signification. Having diagnosed that

> 3.324 In this way the most fundamental confusions are easily produced,

Wittgenstein recommends that

> 3.325 In order to avoid such errors we must make use of a sign language that excludes them by not using the same kind of signs for different kinds of symbols and by not using in a superficially similar way signs that have different modes of signification. That is to say, a sign language that is governed by logical grammar - by logical syntax. (The concept-script of Frege and Russell is such a language, though, it is true, it fails to exclude all mistakes.)

It has been noted that the signification of a term, irrespective of the fact whether it goes as an argument or as a function, is dependent upon the principle of individuation. Then, because 'rose' and 'red', or 'Socrates' and 'bald' are individuated by different principles, they will have different modes of signification; hence they cannot be used as substituends of the same kind of bound variables. This is why in our theories, whose logical structure is exhibited by the standard quantification theory, we have different kinds of primitive terms.

As we find configured in a sentence, be it generalized or not, at least two terms, one constituting the argument part and the other consisting of the function part, with different modes of signification, we will have to treat the sentence as composite. This compositness needs to be analyzed before we can discern the general form of sentences. But as Wittgenstein puts it

> 3.25 A sentence has one and only one complete analysis.

If that be the case, there are a lot of surface itches and some serious brain teasers; for instance, what of the so-called relational sentences like 'Socrates is taller than Simmias'! in the first chapter I attended to the Wittgensteinian demand that in an essential notation, this is to say that in a notation that is nothing more than the one which captures the essence of notation as such, this sentence can be transcribed as

```
SOsi
```

where the relation which is said to be holding between the individual object Socrates and the individual object Simmias, is shown forth by the juxtaposition of the names of the objects themselves, perhaps as in the hieroglyphic sentence above[40]. Then should we treat the form which this sentence shows forth to be real, that is the deep or the logical form of a sentence? Should we treat those two names, namely 'SO' and 'si' to be having two different modes of signification! If anything has emerged from the foregoing it is that Wittgenstein would consider such a language to be rather elementary, to be of any <u>use</u> to <u>us</u>, and that it has too restricted an expressive power to be of any help to us in giving us the type or the kind of descriptions of the world which we intend to give, and do in fact give. (This argument may be seen in the light of what I said about natural history in the previous chapter.) So any language in which different terms of a relational sentence are required to signify in different modes will not be of any use to us, because, among other things, arithmetic cannot be stated in such a language[41]. Should we then consider the form of relational sentences to be

(Ex,y,R) Rxy

or else

(Ex,Ry)(Rx)y ?

Here we face a serious problem, namely which of the two programs should we pursue. Handling this issue, however, requires a separate enquiry; and I shall not attempt at it here[42]. I shall be content with pointing out that Wittgenstein's program is yet to be completed. There are areas of human enquiry where providing an answer is more important than posing a problem; and there are also areas of enquiry where the case is the other way round. Philosophical investigations belong to the latter sort of enquiries. So I have the satisfaction that my effort has issued in a problem. Not that Wittgenstein was not aware of this, for he despaired, in his preface to the Tractatus, of his inability to pursue to their respective logical ends all the perennials he has dug up in the philosophical garden. He writes there:

> "Here I am conscious of having fallen a long way short of what is possible simply because my powers are too slight for the accomplishment of the task. May others come to it later?"

Yet, he succeeded in throwing much light on the general form of sentences by showing that it is the same as the form of a completely generalized sentence, and also by drawing our attention to the fact that names in ungeneralized sentences behave like bound variables in generalized sentences[43]. In trying to highlight these points in the foregoing I continuously used, following the practice of Wittgenstein, only existential quantifiers. This should not induce one to believe that names behave like only existentially bound variables[44]. The choice of quantifier is immaterial for it is a trivial choice. Given the interdefinability of the quantifiers, what I have said can be said using only the universal quantifier in illustrating the point. Wittgenstein opted for the existential quantifier in the Tractatus because

> 5.5321 ... for example, instead of '(x) : fx → x=a' we write '(Ex).fx → fa : (Ex,y).fx.fy'. And the sentence 'only one x satisfies f(x)' will read '(Ex).fx : → (Ex,y).fx.fy'.

Thus if we can give a description of the world by using only generalized sentences, should we drop using names altogether? In reply we might say that though chemistry, we are told, is reducible to physics, there still is a point in doing it the way it is being done all these years. Similarly, there is still a point in using names; only our use of names from now on will be backed by reflection on what such an activity amounts to. We now have a natural history of naming.

Epilogue

Socrates is reported to have said that an unexamined life is not worth living. By this he could not have meant that animals have no right to life; what he perhaps meant was that such a life somehow falls short of being human. Similarly our unexamined use of names has been with us either as a habit or as a ritual; and this is essentially a human habit, for Wittgenstein would say: you may train your dog to respond to an utterance of your name, but you cannot legitimately say that your dog, in so responding, has used your name. How can we let such an essentially human activity go unexamined!

Notes

[1] Incidentally, Wittgenstein thought that this was a meaningless expression.

[2] Vienna Circle.

[3] See Rao (2).

[4] See Rao (7).

[5] I will consider these objections in some detail in the next chapter.

[6] Notation is slightly altered keeping in view typographical convenience, and 'Satz', sticking to the spirit of the conclusions of the first chapter, is translated as 'sentence'.

[7] That is why Wittgenstein's criticism of Russell's theory of identity is at the same time his criticism of Russell's quantification theory.

[8] Obviously, variable-types, and not variable-tokens.

[9] It may be noted that in the universe under consideration there are two books but four copies; as our universe of discourse here consists of copies, by 'book', in this context, we mean 'copy of a book'. How we individuate is an epistemological issue, and how we refer to the individuated is a semantic problem.

[10] For purpose of individuation, "ideologists" need the principle of identity of indiscernibles; but in virtue of the fact that some of them accept or subscribe to model theoretic view, individuation is not required as it assumes that objects come individuated; yet, most of them accept the principle of identity of indiscernibles. Because of these apparent anomalies, too many problems come to the fore, as can be seen from the next chapter.

[11] See the first chapter.

[12] For typographical convenience I shall use 'I' in the same sense as the iota in the Principia.

[13] See Rao (4); nor does the strategy which Quine subsequently designed solve the problem, though it has the virtue of not violating this distinction. Quine wanted to hook uniqueness by introducing singular predicates (instead of singular terms), for these predicates are to be mapped onto unit subsets of the domain of interpretation while allowing the variables range over the entire domain. The relationship between these singular predicates, for example 'Socratises' and the object of which it is true, namely the individual object Socrates, is as intriguing as the relationship between a name and its referent. Quine's strategy constitutes his way of solving the problem of uniqueness while adhering to the referential inclusive interpretation of the variables. There is no reason why we should indulge in this syntactical reformulation, rather than opt for the Wittgensteinian reinterpretation.

[14] What I am saying here needs to be taken together with Foegelin (1), specially Ch.5, Sec.v. Though I differ with him on many issues discussed in the other parts of his book, I treat this part of it pertinent to the issue at hand. Yet, the way he is taking Wittgenstein's interpretation of variables sounds rather misleading. However, I am delighted to see him in

agreement with my 1965 argument in (1). He writes that "writers on Wittgenstein do not usually stress the role of the application or use of language in the 'Tractatus'. Some are different about projecting back upon the 'Tractatus' doctrines thought characteristic of Wittgenstein's later writings. Others seem anxious to maximize the distance between the 'Tractatus' and Wittgenstein's later writings in order to make the transition more dramatic. In fact the notion of application is central to 'Tractatus', for it is only through application of language that we are able to resolve many questions that defy proper formulation within our language."

However, I disagree with Foegelin on another point which he raises in that chapter. It is not the case that "the 'Tractatus'", as he thinks, "is silent about the way we picture the world using only fully generalized propositions." We find Wittgenstein "agonized over this problem" not only in Note Books, as Foegelin notes, but also in the Tractatus which he seems to be ignoring. That this is the case should be evident from the next few pages.

[15] Wittgenstein's objections to a reluctance towards extending quantification theory to comprehend identity could be treated as his refusal to transfer what belongs to semantics to syntax. Similarly, his objections to Russell's theory of definite descriptions could be considered as his reluctance to smuggle in what belongs to pragmatics to syntax. This way of looking at these matters fits in with the general spirit of his vision of language, namely that all that is there to a language cannot exhaustively be captured in any specification of its syntax, however, elaborate this specification might be.

[16] Or as it was cryptically put in Rao (1), a picture is a picture because it is used as a picture.

[17] As this sort of a description of the world is geared to represent the world uniquely-1, what Wittgenstein is suggesting may be likened to what Marx is attempting at in the Capital. Marx is trying to give a description of the capitalist economy in terms of the total social capital and total social labour etc., which are a bit too much ethereal and yet are related to concrete social reality. Marx is trying to give an abstract representation of the concrete; and it should be differentiated from concrete representations of the abstract.

[18] Emphasis added.

[19] Another reason for this is that such a move may be possible of the world is finite. But as Wittgenstein would say, the assertion that the world has a certain number of objects is devoid of sense.

[20] This trick was perhaps borrowed by Russell from Moore who is adept at it. The one exception to this normal practice of the then Cambridge dons is Keynes. See his preface to his General Theory; he writes there: "I myself held with conviction for many years the theories which I now attack." This could perhaps be due to the reason that economists have better memories and philosophers have bitter memories.

[21] Emphasis added; even Ramsey is a party to this practice. That the all-sentences are products and that the some-sentences are sums is a Cambridge confusion which Wittgenstein could not stomach. There are too many entries in his Note Books where he attacks this view and gives his own. But it was too deeply rooted in Cambridge philosophy. Ramsey for instance writes in Foundations, pp.54-55: "I adopt the view of Mr. Wittgenstein (and also, apparently, of Mr. Johnson ...) that 'for all x, fx' is to be regarded as equivalent to the logical product of all the values of 'fx', i.e. to the combination of fx_1, fx_2, and fx_3 and ..." He repeats the same on p.159.

[22] Emphasis added; it is on the basis of this criterion I argued that Wittgenstein should have objected to Russell's theory of descriptions on the ground that it homogenizes expressions. Methodologically Russell's discovery that (Ix)Ax and its well-known definitional equivalent say the same, and Wittgenstein's discovery that 'fa' and (Ex). fx. x = a say the same are at par. Both of them are, to use Ramsey's phrase, paradigms of philosophical analysis in virtue of showing how the apparent structure of the definiendum is deceptive and how its real structure is the one exhibited by the definiens.

[23] See for instance Quine's Methods; it could be a fruitful exercise to go to it after going through the Tractatus. There is nothing wrong in our paying attention to the use of constants a and b; Wittgenstein himself remarked in the Tractatus: "What a sign does not show, its application reveals."

[24] If 'everything is momentary' is true, 'the flower that blossoms in the morning fades away in the evening' is also true, for the former (a slogan of the Indian Buddhists) is as much about you and me as it is about the flower; the latter is the way in which the Chinese Buddhists translate it! Perhaps that is why each of us has some fragrance! Incidentally, does this show that the Chinese mind is more concrete, or that the Indian mind is more abstract!

[25] See the Appendix to his (10).

[26] I may briefly betray my notion of an individual without either imputing it to Wittgenstein, or arguing it out in detail. It is classical and oriental which can be summed up as: an individual is unique (unprecedented), does persist, is inexhaustible, and discloses itself.

[27] Recorded in his Vienna Circle.

[28] This obviously includes the Note Books period.

[29] The third issue which is epistemological and pragmatic, namely how objects are individuated, has been harped on sufficiently to be mentioned again.

[30] Russell's definition of a number is not the way in which only abstract objects are individuated; it is neither unique to abstract objects, nor is idiosyncratic of the Russellian way of doing things. It is one of the ways in which any object can be individuated. To note how this seems to have been used by Aristotle, see Rao (2).

[31] I am not saying that this mass is the out-put of the functioning of the apparatus; hence there is no need to activate the anti-idealist guns.

[32] Recognizing this may go some way to solve at least some of the riddles in Frege-Russell philosophy of language.

[33] This might be the legacy of Aristotle; see Rao (2). Obviously, if a theory is intended to be about a heterogeneous set of objects, then it will have to have as many principles of individuation as there are types of objects in the domain.

[34] Each discipline is an attempt to rewrite a bit of the Genesis, astronomy, geology, zoology for example. Why not philosophers have a share in this blasphemous activity!

[35] The object which figures in an ostensive definition is what Wittgenstein would call 'the bearer of a name' and is different from the referent of a name.

[36] This dependence is not to be interpreted idealistically in the sense that it amounts to holding that a theory generates its own reference; that would be patently Hegelian and goes against the Wittgensteinian realism.

[37] Until and unless the principle of individuation of properties is settled for, it cannot be said whether Leibnitz's law of identity, as it is usually understood, this is to say that in the sense in which numerical identity and qualitative identity get telescoped, holds or not.

[38] This should be evident from a more familiar experience; in certain descriptions we treat a property simple, and in others we treat the same as complex.

[39] Notation slightly altered, and emphasis added. From the occurrence of 'independently' here, we can infer that Wittgenstein is rejecting both the bundle-theory, this is to say that the view that an object is nothing but a bundle of properties, as well as bare particulars, which Russell dismissed contemptuously treating them as coat-hangers. Is it the case that Wittgenstein, like Aristotle, is suggesting that neither objects nor properties come in isolation! That fits in with his facts-as-the-logical-discernibilia view of the world. Did not Aristotle, in Posterior Analystics, say that Socrates, for instance, derives his nature by being a man, this is to say that by belonging to a genus.

[40] As early as October 1914, see Note Books, Wittgenstein realized:

> "The difficulty of my theory of logical portrayal was that of finding a connection between the signs on paper and a situation outside in the world."

This is the difficulty that we face in projecting so to say the "multi-dimensional relations" which obtain in the world into a two-dimensional script to which we happen to be accustomed. The problem of equinumerosity of a relational sentence and the relational fact can be located here. As a word of caution I may add that, as sentences and facts as such are relational in their nature, the previous sentence should be taken as something which is framed in that shape only in a manner of speaking.

[41] In this connection we need to note two things; (i) this will be the case even if we banish Platonism interpreting arithmetic as Wittgenstein does, and (ii) when n and m are natural numbers, imagine how unimaginably complex doing arithmetic turns out to be if we need to individuate the referents of n and m respectively in say n+m by different modes of individuation. Compare it with the way in which we usually individuate natural numbers with a simple recursive definition.

[42] I am inclined towards (b); see Rao (3).

[43] As Foegelin remarks, Wittgenstein was perhaps "agonized" over the nature of generalized sentences. It was evident from the Note Books that he was engaged in an intense philosophical activity concerning it throughout October 1914. Some of his thoughts of that month as they were recorded there were subsequently abandoned; one of these is that logical connectives have referents. Some others were reformulated to go into the Tractatus, for instance his view on generalized sentences. At the beginning of October he was still wondering whether generalized sentences were a priori (!) and hence were either analytically

true or analytically false. But having noted, during the middle of the month, that [(Eϕ)(x)ϕx] is neither a tautology nor a contradiction, he abandoned that idea and came to the view which we find in the Tractatus. Central to this view are the ideas that (a) they are structural descriptions of the world, (b) it is not the case that they have structures, but that they are structures, and (c) in virtue of being descriptions of the world they can be either true or false. To cite from the Note Books,

> 17.10.14 ... it is possible to devise a picture of the world without saying what is representation of what.
> 28.10.14 What the completely general sentences describe are indeed in a certain sense structural properties of the world. Nevertheless these sentences can still be true or false.
> 29.10.14 (The subject matter of general sentences is really the world; which makes its appearances in them by means of a logical description ...)

[44] There is a likelihood of even brilliant minds embracing such errors. Did not Church think that existential commitment creeps into quantification theory only through existential quantifiers; but this is a classic case of a Genius' inadvertency.

Chapter Four : On Identity

Wittgenstein's theses, remarks, investigations, ...[1] all are indicative of his life-long struggle to get rid of the puzzlement over identity. With the exception of the more general problem concerning representation of the world in language, no other issue preoccupied him that long. We find him writing to Russell as early as October 1913 informing him how it made him philosophically restless, and we see him still engaged with it in (3) published soon after his death in 1951. In that letter, he writes:

> "Identity is the very devil immensely important, very much more so than I thought. It hands, like everything else, directly together with the most fundamental questions concerning the recurrence of the same argument in different places of a function. I have all sorts of ideas for a solution of the problem but could not arrive at anything definite. However I do not lose courage and go on thinking."

From the first sentence of this passage it should be clear that some time earlier than 1913, thus almost as soon as he switched over to philosophy, Wittgenstein must have experienced his puzzlement, and it is certain that he must have experienced it while going through the works of Frege and Russell[2].

By the time he came to write (4) in 1892, Frege was convinced that "identity gives rise to challenging questions which are not altogether easy to answer". He lists three such issues, namely (a) is identity a relation? (b) if it is a relation, is it a relation between objects or symbols? And (c) are not a = b, and a = a the same when the former is true? These are nested issues; an affirmative answer to (a) is needed for (b) to make sense, and an answer to (b) is required to make sense out of (c). Frege, however, does not handle (a) and (b); on the other hand he works out that paper as if he posed (c) in isolation and was attempting to solve it[3]. Wittgenstein was the first, at least one among the firsts, to be puzzled about this[4]. The clearly incompatible and apparently incoherent statements that Frege makes about identity must have intellectually irritated him. For instance Frege admits that though to our primitive intuition a=a and a=b appear to be conveying different things,

> "(I)f the sign 'a' is distinguished from the sign 'b' only as an object and not as a sign, the cognitive value of a = a becomes essentially equal to that of a = b, provided a = b is true."

As he suggests that we then better treat identity, this is to say that the sign of identity, as something which occurs along with other signs. What is surprising is that he does this in complete innocence of what he has been doing since 1979, and also, should we say, in total indifference to what happens to his "formal language of pure thought" and his concept-script[5] when identity is treated in that way. Worse is that he treats identity as a relation, of course, sometimes as a relation between objects, and sometimes as a relation between signs. Eleven years after (4), in his (3) Frege was explicit that

> "... identity is a relation given to us in such a specific form ...",

and thirteen years before that in (1) he treated identity as a relation between objects except in the case of its occurrence in a definition where it is said to be a relation between signs. To account for this dual-functioning of identity, Frege proposed in his introduction to (1) a theory according to which names[6] refer in two different ways; first they refer to objects, and secondly they refer to themselves. But he relegates all this to an inefficacious background when he comes to develop his concept-script which from now on shall be referred to as the Standard Quantification theory (SQ, for short)[7]. Yet, it is rather surprising that at no time since 1879, Frege found fault with SQ; nor did he show any awareness of the possible anomalies which would creep into SQ when identity is allowed to figure with the dual-function in question. Throughout his intellectual career he held that the logical structure of language coincides, or be made to coincide, with SQ; and it is no news to him that SQ requires identity to be a relation, a relation between objects for that matter. He knew it well that the identity predicate will have to be taken as a two-place predicate occurring between names referring to the members of the domain of interpretation[8].

A definition of identity (the identity axioms are expected to provide a complete implicit definition) if it is to be plausible at all, should provide for reflexivity, symmetry, transitivity, and substitutivity so that it may serve its intended purpose. In SQ transitivity is available because it either occurs among its connectives or else is definable in it; in any case it partially orders its set of theorems. So only reflexivity and substitutivity are axiomatically assumed and the symmetry of the identicals is deduced[9]. Let us note what happens to the deduction of symmetry when we refrain from treating identity as occurring between signs referring to objects, or variantly when we treat the signs between which it occurs as referring to themselves. Evidently, in the axiom covering substitutivity, this is to say that in

(i) $x = y \rightarrow (\phi x \rightarrow \phi y)$,

the same term, namely x, comes out with different referents; x in its occurrence in the antecedent will refer to itself, and its occurrence in the consequent will refer to the members of the domain of interpretation. To put it differently x occurs (i) autonymously as well as heteronymously. Now given reflexivity, this is to say that given

(ii) $x = x$

in order to arrive at

(iii) $x = y \rightarrow y = x$

through (i), we need

(iv) $x = y \rightarrow (x = x \rightarrow y = x)$.

(iv) is available only through a substitution of $x = x$ for ϕx, or of $x =$ for ϕ [10]. But such a substitution is not possible, nor is it permissible on the count of our intuitive rules of inference because in $x = x$, x refers in one way and it refers in another way in ϕx. This makes the whole inference anomalous. But Frege, Russell, and Wittgenstein all shared that SQ, with its referential interpretation, captures the essence of the syntactical and the semantic structures of our language of representation. Roger White thinks that by trusting the story and ignoring the story-teller, thus taking into account what Frege does in the text (1) rather than what he says in the introduction to it, we can surmise how he got rid of this annoyance; and in the text Frege did take identity as a relation between objects. But to be fair to relations, Wittgenstein seems to have wondered as to what sort of a relation identity could be; and what after all is being related in $x = x$! As he came to conclude on 27.10.1914,

> "' $x = x$' is not a sentential form. (Consequences)."

Then why not reformulate SQ such that the resultant theory would be free from pseudo-relations like identity? After all, Frege did treat 'existence' as a second order predicate, and obviously existence and identity go together like Juno's Swans! So why not treat identity too as a second order predicate? That seems to be one half of Wittgenstein's problematic. The other half perhaps became clear to him while going through Russell.

Russell too, like most philosophers after him, took into his cognizance what Frege did, but unfortunately what he did in (4) and thus got trapped in epistemological issues at the cost of the semantic problem. He thus came to see the issue about the cognitivity content of identity sentences as the issue - the problematic, that is. As he thought

> "(i)dentity is rather a puzzling thing at first sight ... (W)hen you say 'Scott is the author of Waverley' you are half tempted to think there are two people, one of whom is Scott and the other the author of Waverley and they happen to be the same. That is obviously absurd, but that is the sort of the way one is always tempted to deal with identity."[11]

If Frege's failings have only puzzled Wittgenstein, Russell's fusion and confusion must have distressed him. Paying exclusive attention to (4) where Frege does not even mention the dual role of names - that is their occurring as names of objects and as names of themselves - Russell accepted the twin-function semantics of Frege; he subscribed to the Fregean belief that a name has a referent as well as a sense which is too prominent and all pervading in that article. Added to this costly oversight is Russell's habitual strategy of imputing his own earlier views to others and then attacking them to overcome his own earlier belief. He imputed to Frege the view that names are abbreviations of descriptions, something that he himself held for some years[12]. Worse is Russell's absorption of the Fregean distinction between sense and reference into English philosophical lingo by using the Millian 'connotation' and 'denotation' with their epistemological and psychological affiliations. Frege held a very broad conception of a name which turns proper names and definite descriptions semantically homogeneous; and according to this conception 'Scott' and 'the author of Waverley', 'the current inflation' all are names. It is Russell who subscribed to an emaciated notion of a name which Quine subsequently collapsed into that of a bound variable. Anyhow, having accepted the truth-function semantics of Frege, having rejected the view of names which he imputed to Frege and having subscribed to a very narrow conception of a name, Russell had to split his puzzlement about the informativeness of identity sentences into several small puzzlettes, for instance puzzlettes about (1) n = n,

(2) n = d, (3) n_1 = d, (4) n = n_1 and (5) d = d_1 - where n and n_1 are names and d and d_1 are definite descriptions. What I need for my purpose here, which is to rip open the conceptual belly of the Principia view of identity and see where the entrails got entwined, are identity sentences of the sort (3) and (4); for this limited aim it would be sufficient to take into account sentences of the form 'a is b' and 'a is the F' where the copula is understood to be expressing identity, and understand those sentences against the background of Russell's theory of definite descriptions and also his views on proper names. The logical structure of 'a is the F' on the strength of that theory turns out to be

(I) $(Ex)\{Fx.(y)(Fy \rightarrow x = y).a = x\}$.

A glance at (I) is all is needed to note that in it identity is taken precisely in the Fregan sense, namely as a relation between objects. But this is only part of the story; the rest of it may be noted by paying attention to the fact that in (I) the identity sign occurs twice, first as an integral part of the quantificational apparatus, and secondly as a part of the content of the sentence that (I) is supposed to be symbolizing. Roger White is of the view that this equivocal occurrence of the sign of identity in (I) must have been noticed by Wittgenstein, and also that he must have thought that this is a bit circular in the sense that if the theory of descriptions is intended to throw some light on the notion of identity, specially on the informativeness of the identity sentences - identity itself should not have figured in the quantificational apparatus, because it is being used here as a tool in the analysis. Yet, if it did figure as it did, this could only be due to the reason that, as Wittgenstein notices, quantificational mechanism and identity are inextricably linked together.

White credits Wittgenstein of having noticed two things in Russell's analysis of sentences of the form 'a = the F', this is to say that in the analysis in which his views on identity get telescoped with his views on the semantic behaviour of the definite descriptive phrases[13]. As can be gathered from the

Tractatus 5.531 - 5.5321, Wittgenstein noticed that (I) has simpler equivalents in, for example

Ia. $(y)(Fy \rightarrow y = a).Fa.$

and

Ib. $Fa.(Ex)(y)\{Fx.(Fy \rightarrow x = y)\}.$

But such simple equivalents are forthcoming only in the case of = and not in the case of any other two places predicate R in 'a R the F'; and they are available in the case of = because of its dual role which lurks out in the contexts of descriptions and definitions. Unlike (I), in (Ia) and (Ib) the identity sign occurs between bound variables, thus as an integral part of the quantificational mechanism. As such it cannot be treated as a relational expression. It is likely that Wittgenstein noted this quite early in his enquiries into the nature of identity[14] for we find him writing to Russell in December 1913:

> "... your "Theory of Descriptions" is quite certainly correct, even though the individual primitive signs in it are not at all the ones you thought."[15]

Further, in 5.5301 of the Tractatus he remarks:

> "It is self-evident that identity is not a relation between objects. This becomes very clear if one considers, for example, the sentence '(x):Fx. x = a'. What this sentence says is simply that only a satisfies the function F, and not that only things that have a certain relation to a satisfy the function. Of course, it might then be said that only a did have this relation to a; but in order to express that, we need the identity sign itself."

All this, however, is Wittgenstein's comment on the Fregan legacy in Russell. Btu Russell has his own misunderstanding of the nature of identity, and this in addition to his misunderstanding of Wittgenstein's point of view concerning identity. Wittgenstein responded to both these misunderstandings. In presenting Russell's misunderstandings and Wittgenstein's responses to them I shall take into account the central point in the results of Muehlmann's research, briefly present it, further elaborate the important points which he seems to have ignored, and also reject a part of his otherwise interesting and subtle argument. His central thesis is that there are three distinct senses of identity, that Russell continuously confuses these senses, and that Wittgenstein notices this confusion. When we say that x and y are identical, it might mean that they are numerically the same, or that they share the same qualities, or that they share the same qualities <u>and</u> relations. Let us use three different signs, namely $=_n$, $=_q$, and $=_p$ corresponding to these three different senses respectively. The last is supposed to attract the attention of only philosophers. Wittgenstein finds $=_p$ incoherent, and $=_q$ useless for representational purposes because it involves collective reference. This leaves only $=_n$; and this is treated as a genuine notion. What Wittgenstein was after was capturing the nature of identity in the numerical sense of the term. As an example of the confusion between the different senses of identity in the writings of Russell, Muehlmann cites the following passage from the <u>Principles</u> with which Wittgenstein was quite familiar:

> "The question whether identity is or is not a relation, and even whether there is such a concept at all, is not easy to answer[16]. For it may be said, identity is not a relation, since were it truly asserted, we have only one term whereas two terms are required for a relation ... Nevertheless identity must be something. We might attempt to remove identity from terms to relations, and say that two terms are identical in some respect when they gave a given relation to a given term. But then we shall have to hold either that there is a strict identity between the two cases of the given relation, or that the two cases have identity in the sense of having a given relation to a given term; but the later view leads to an endless process of the illegitimate kind. Thus identity must be admitted, and the difficulty as to the two terms of a relation must be met by a sheer denial that two different terms are necessary. There must always be a referent

> and a relatum, but these need not be distinct; and where identity is affirmed they are not so."[17]

Muehlmann points out that even if the absurdity of the contents of the passage is ignored, it is impossible to miss - and hence Wittgenstein could not have missed - the shifting meaning of the term 'identity' in its disparate occurrences in the passage. In the first and the second sentences what is at issue is numerical identity; what figures in the fourth sentence is the philosophers' concept of identity. Finally in the last sentences there is reversion to numerical identity. When this is juxtaposed with what Russell has to say in the Principia, the picture becomes more confusing. = as it occurs in x = y in its occurrences in the Principia is not even numerical identity; it is the same sign of equality in mathematics, for example as in 3 + 2 = 5. As such it cannot be the case that it is the sense of identity that Russell is trying to capture in the so-called definition of identity in the Principia which reads as follows:

> "the propositional function 'x is identical with y' will be written as 'x = y' ... and is defined as follows:
> 13.01 $x = y :\equiv: (x) : \phi!x \rightarrow \phi!y$ Df.
> This definition states that x and y are to be called identical when every propositional function satisfied by x is also satisfied by y."[18]

Whereas Wittgenstein seems to have had more important reasons for being dissatisfied with this definition, it has come to be thought that Wittgenstein's attitude towards it was influenced by his recognition that 'identical' as it occurs in Russell's introduction to the Tractatus, specially the sentence 'if every property of x is a property of y, then x is identical with y' - cannot be taken to be having the sense it receives in 13.01. This is so because otherwise 13.01 would turn out to be a tautology and also because Russell does not seem to take it as a tautology. This argument may gain some force if 'tautology' is replaced by 'analytic'; but that is the point in which we should not be interested in. What is of interest is that this argument is based on a (literally chronic) error. Russell's introduction to the Tractatus was written a decade later than his formulation of 13.01, and the Tractatus in which Wittgenstein explicitly rejected 13.01 was written three years before the introduction. Yet, there is nothing wrong in highlighting the confusion between the principle of the dissimilarity of the numerically distinct on the one hand, and the principle of identity of the indiscernibles on the other. But it should be kept in mind that this is not idiosyncratic of Russell; he is not alone in confusing them. Because that confusion is required for reducing numerical identity to qualitative identity, or qualitative identity to numerical identity, too many philosophers voluntarily embraced it. But given such a confusion of the two principles, definitions like 13.01 turn out to be tautologies (at least in the sense of the term in which it is used as a synonym for being analytic.) According to Muehlmann, Russell is left with only one way out for being able to secure some content to his definition. It is to treat = as it occurs in 13.01 as a primitive predicate; and he thinks that it is what Russell is doing, "unknowingly though".

It was argued that as 13.01 does not throw much light on identity, it would be better to pay attention to what Russell says about it elsewhere in the Principia, specially on p.57 of Vol.1, namely

> "... we cannot, without the help of the axiom (of reducibility) be sure that x and y are identical if they have the same predicates. Leibnitz's identity of indiscernibles supplied the axiom. It should be observed that by "indiscernibles" he cannot have meant two objects which agree as to all of their predicates, for if one of the properties of x is to be identical with x, and therefore this property would necessarily belong to y if x and y agree in all their properties."[19]

But, as Russell realized, this is impossible if they are two objects. However, even this is not of much importance for our purpose here. What is worth noting is that Russell is aware of the fact that 13.01 cannot be used for its intended purpose without the principle of the identity of indiscernibles. Thus he came to see that he is obliged to guarantee the logical necessity of the principle of identity of indiscernibles. What Wittgenstein must have noticed is that Russell failed in meeting this obligation. But Russell in his turn took this as Wittgenstein's rejection of the principle itself. In his attempt to secure logical necessity to the principle, Russell committed, in Wittgenstein's view, a second sin: he treated diversity too as a relation[20]. His first sin obviously was treating identity as a relation. While arguing why Leibnitz must consider the principle as necessary, Russell writes that

> "... suppose A and B were two indiscernible substances. ... A differs from B, in the sense that they are different substances; but to be thus different is to have a relation to B. This relation must have a corresponding predicate of A. But since B does not differ from itself, B cannot have the same predicate. Hence A and B differ as to their predicates, contrary to hypothesis."[21]

In fact it his gamble for securing certainty to that principle that pushed Russell to that famous - but as Wittgenstein felt, rather infamous - axiom of reducibility. Russell ran into difficulties in defining identity along the Fregean lines, largely because of the type restrictions that he accepted. Suppose that $x = y$ is defined as $(\phi)(\phi ! x \rightarrow \phi ! y)$, since the variable ϕ is restricted to an order, the definition does not yield the desired properties of identity. But given the axiom of reducibility, within the framework of the Principia, 13.01 can be used in the intended way because if $a = b$, and ϕa, then whatever might be the order of $\phi ! x$, one can invoke the axiom to infer, for some predicative function $\phi ! x$, if ϕa, and $a = b$, then ϕb. And, if the principle is taken as Russell did, this is to say that as an assertion of

$$x \neq y . \rightarrow . (E\phi)(\phi ! x \, . \sim \phi ! y).$$

13.01 can be used if, instead of the axiom, the principle itself is used; since this formulation of the principle is provable in the Principia, of course using the axiom, Russell came to hold that the axiom is a "generalized form" of the principle as he formulated it[22]. In a work whose composition coincided with the publication of the Tractatus, Russell writes:

> "... identity of indiscernibles will state that if x and y agree as to all their predicates, they are identical. This can be proved if we assume the axiom of Reducibility. For in that case every property belongs to the same collection of objects as defined by some predicate. Hence there is some predicate common and peculiar to the objects which are identical with x. This predicate belongs to x since x is identical with itself, hence it belongs to y, since y has all the predicates of x, hence y is identical with x."[23]

Thus if the logical roots of 13.01 are in the axiom, Wittgenstein's dissatisfaction with it is rooted in his unhappiness with that axiom. It will not be far from truth to say that Wittgenstein moved into Cambridge with a bagful of objections to that axiom[24]. As early as June 1913 he wrote to Russell:

> "The whole axiom seems to me at present a juggling trick. Do let me know if there is more to it."

Five months later he wrote again stating emphatically that

> "it is not a logical proposition at all."[25]

Explicating and exploiting what is embedded in the framework of the Tractatus, we can give several arguments in support of Wittgenstein's attitude towards that axiom. One such argument, perhaps the least technical one, would be that a world in which the axiom will hold may be possible, but such a world will not be our world because the structure shown forth by that axiom will not be isomorphic with anything that obtains in or with our world; it will not hold with respect to our world. Though what is can as well be otherwise, the logical is supposed to be invariant under all situations. Thus Russell's definition is without the logical force which he claimed for it, and which he needs for it. 13.01 is not satisfactory on other counts too. Another source of irritation is that what is hoped to achieve by it cannot be achieved unless reflexive predicates, negative predicates, are similar unnatural predicates - unnatural in virtue of containing referential occurrences of individual terms - are allowed to go as substituends for ϕ. This is the same as to say that what Russell means by the variable ϕ includes not only predicates in the Leibnitzian sense (to be precise, in the sense in which Russell took Leibnitz to be understanding that term), but also what he himself calls predicate functions - for example, 'not identical with Socrates', 'identical with Plato', '$\neq$ x', and '= y'. But this strategy is required to turn the principle into a necessary principle; the passage above can be treated as Russell's confession to that effect. This means that what Russell is defining, as he himself seems to have thought, is at most numerical identity; and he is defining it in terms of the philosophers' sense of identity[26]. Given the principle, if apparently two objects are identical in the sense of $=_P$, then they are really one. As he says in the Principia

> "... identity ... can only hold between x and y if x and y are different symbols for the same object."[27]

Note the marriage of the semantic and the ontic; but that union, holy or unholy, is needed by Russell because what he is after is a substitutivity criterion without which the extensionality thesis to which he was deeply committed will not hold; and without that thesis the whole program of the logistic could collapse. (It collapsed for a different reason, but that should not detain us here.)

It should be clear now that too many things are amiss with 13.01; but from none of these it follows that the definition proposed in it is inadequate; Wittgenstein thought that it was inadequate too. That it is so follows from the stand Wittgenstein took on the axiom and the principle. If in 13.01 Russell is defining numerical identity in terms of qualitative identity, the definition would be adequate only if the principle is true, because it is only that principle that can guarantee that qualitative identity entails numerical identity. Otherwise the left side of 13.01 could be false and the right side of it true. This could be the import of the parenthetical remark in 5.5302 of the Tractatus. If 13.01 is inadequate, it is so in this respect; this is the same to say that it is insufficiently sustained[28].

In view of the fact that Wittgenstein subscribed to the Frege-Russell belief that SQ is a theory of reference, and that it shows in essence the structure of the representational language as such, his "rejection" of the Fregean and the Russellian theories of identity leads to a new problematic; this is so because without being extended to envelop identity, SQ as developed by Frege and Russell falls short of its expectations - specially in capturing uniqueness. What then is to be done to SQ so that it may be adequate for the intended purpose when the Fregean and the Russellian routes are avoided! The issue for all the three, Frege, Russell, and Wittgenstein, was about making available an anomaly-free formulation of SQ which is adequate for representation; only what Wittgenstein saw to be clearly anomalous was not taken to be so by the other two. If Wittgenstein thought that Frege-Russell extension of SQ is anomalous, it was because he was more respectful of the ordinary language than they were; he was respectful to it even during the Tractatus, contrary to the popular belief about him. Thus he was more informed by it, and had better intuitions into its structure. Frege, for instance, showed scant respect to ordinary language; he was even condescending towards it as it can be noticed from the following remark which can be found on p.67 of his (6):

> "It cannot be the task of the logician to investigate language and determine what is contained in a linguistic expression. Someone who wants to learn logic from language is like an adult who wants to learn how to think from a child ... Languages are not made so as to match logic's ruler."

From a child! Why not? Our theories about children, as Wittgenstein would argue subsequently, may be childish, but that does not mean that children think childishly. In any case his respect for ordinary language led Wittgenstein to notice that the natural way of reading quantifiers and the normal use of definite descriptions do not involve the notion of identity, not to say anything further about identity as a relational part of the definite descriptive phrases. For several F, on occasions, we do say that 'there is at most one F' but at no time we seem to mean by that, that any two F's stand in the relation of identity to one another.

While paying attention to the role of identity in connection with Russell's theory of descriptions, we noted how in (I) above it figures with different uses; for convenience let us call it in its different uses formal identity and content identity. In (Ia) and (Ib) which are simpler equivalents to (I) it occurs only as formal identity; it occurs only within the scope of the quantifiers. This use of the identity sign cannot be explained in terms of other relational terms, and all other uses of identity need to be understood or explained in terms of this use. But this use itself belongs to the quantificational apparatus. Unlike in Begriffsschrift and Principia, in the concept-script identity then should be represented as a part of the quantificational mechanism, and not a separate relation. In 5.5301 and 5.5321 of the Tractatus Wittgenstein is suggesting ways towards such a representation[29]. His suggestion is so simple that it has far reaching consequences on our understanding of representation and language themselves. That for Wittgenstein philosophy is a Sprachcritik in the precise sense of being a description of its deep structure or syntax is a commonplace. But what is not noted by several students of Wittgenstein, at least with precision, is that it is not confined to a critique of the natural language, but covered also the syntax of the formal languages which were developed with the intention of capturing the essential syntax of the natural languages. Wittgenstein thought of his critique in contra-distinction with the critiques of Mauthner, Krauss, and Freud; all these are semantic in some sense or the other, this is to say that they are alternative mixes of the political, the poetic and the psychological. When this fact is taken into cognizance, it becomes clear that

Wittgenstein's critique is geared towards syntax. Again that the SQ was intended to capture the essence of syntax is a commonplace; but it is not sufficiently appreciated that Wittgenstein's most important contribution to philosophy - given that philosophy is Sprachcritik, that is - is his critique of the Frege-Russell version of SQ, this is to say that his critique of the syntax of SQ with the intention of setting us free from the deceptions of its syntax. The syntax of Frege and Russell versions of SQ tempts us to take = either as a relation or as a function; and it is neither. Consider for instance the following formulas of their versions of SQ:

(i) $(Ex)(Ey)(Fx \,.\, Fy \,.\, x = y)$, and (ii) $(Ex)(Ey)(Fx \,.\, Fy \,.\, x \neq y)$.

If (i) is true, then it is trivially equivalent to (ia) $(Ex)(Fx \,.\, Fx)$, thus to $(Ex)(Fx)$, and if (ii) is true, then it is equivalent to (iia) $(Ex,y)(Fx \,.\, Fy)$. What was intended to represent in (i) and (ii) can as well be represented in (Ia) and (IIa), and can be represented without using identity-sign, but relying only on the identity of signs. This shows that = is not an indispensable part of the concept-script. In the Frege-Russell version of SQ coincidence of different variables[30] is permitted. Consequently in the concept-script so developed the identity sign becomes an essential requirement for capturing uniqueness. Wittgenstein's point is that such a coincidence be banned, the coincidence of the values of different variables be excluded, and uniqueness be captured without resorting to identity. The difference between these two interpretations of the quantifiers, or these two different readings of the variables, has been brought to the fore by Hintikka[31] in a well chosen example from geometry. Whereas the former, which is called the inclusive interpretation, would permit assertions like

(a) any two points of a straight line completely determine that line, and the latter, which is called the exclusive interpretation requires the same to be asserted as

(b) any two distinct points of a straight line completely determine that line[32].

However, this is neither dispensing with nor transcending the need for the notion of identity. It is only a new way of embedding it in quantification theory, in the concept-script that is. If SQ is so restructured, the resultant theory will be less expensive. Every significant formula of the revised theory finds a translation into the original theory, and every formula of the latter which is significant in the sense which Wittgenstein gave to that term, yields to a translation into the revised system. As I shall show a little later, their respective classes of significant formulas are not the same; this is an important point which Hintikka overlooks and consequently misses the salient feature of the revised system. Hintikka does not

> "quite see the superiority ascribed by Wittgenstein to an exclusive interpretation of variables. An exclusive interpretation of variables can be carried out so that it gives us a new way of saying exactly the same things we could say before. Our experimental non-conformity"

according to him

> "does not enable us to do things we could not do before."

Wittgenstein would certainly have disagreed with Hintikka on this point because he would have seen and shown to others that these two systems do not permit us to <u>say</u> the same things which is what Hintikka believes. In the original system one can say things which on the strength of the new system are meaningless, either <u>sinnlos</u>, or <u>unsinning</u> as the case may be. And there are things which could be said meaningfully in the original system, in so far as the sign of identity is used as a part of the quantificational mechanism and is allowed to figure within the scope of the quantifiers, and are <u>shown</u> in the revised system. Because of this, the new system is more perspicuous, logically as well as ontologically.

Hintikka formulated the translation rules for these two systems. To state them slightly modifying the notation to suit to us, where ($\underline{E}$) and ($\underline{U}$) are exclusively interpreted quantifiers, and (E) and (x) are the standard inclusive quantifiers, and K a formula not containing the sign of identity - and thus a formula of either of the system – these rules are:

($\underline{E}$) $K \equiv (Ex)(x = y_1 \vee x = y_2 \vee \ldots\ldots \vee x = y_k . K)$.

where $y_1, \ldots, y_k$ are all the free variables of $(\underline{E}x)K$, and

($\underline{U}$) $K \equiv (x)(x = y_1 . x = y_2 \ldots x = y_k . K)$.

where $y_1, y_2, \ldots, y_k$ are all the free variables of $(\underline{U}x)K$.

What has been achieved is thus an implicit formalisation of identity as a part of the formalisation of the exclusive quantifiers. Thus,

$$(Ex)(Ey)(Fx . Fy . x \neq y)$$

can now be written as

$$(Ex,y)(Fx,y)$$

and can be assigned the truth-value 'true' only if Fx,y is true for some value of x, and a different value of y. The identity sign does not figure in the concept-script as a diadic relational term as it does in Begriffsschrift and the Principia. Yet it may be represented by a convention concerning the reading of quantifiers, for instance for recognizing that different tokens of a variable-type within the scope of a quantifier receive the same values; 'A' is the same as 'A', as 3.203 of the Tractatus tells us. Of course, it is admitted that it is rather "cumbersome" to use the exclusively interpreted quantifiers; but it was never a part of Wittgenstein's intention that one should be nice to the lazy. His aim was to show first the possibility of an alternative notation, an alternative syntax, an alternative concept-script, and secondly to achieve maximal logical and ontological perspicuity. Once this is taken into account we can revert back to our old ways of doing things, may be now with a few cautionary conventions. It may be thought that in leaving the argument at that point I am trying to be conciliatory to Hintikka. I admit that I am, but not because I accept what he said about the Wittgensteinian alternative. As I see it, pushing further my objections to him is not pertinent to the point at issue. Yet I may indicate the line on which the argument can be extended to a point where it goes against what Hintikka has to say. His argument assumes that logic is philosophically neutral. This may not be the case; and what might be true is that philosophical positions differ in what they respectively take logic to be - what they take as a logic. Thus, for all that we know, it could be that through logical differences philosophical differences become manifest or come to the fore!

Hintikka's claim that every formula of one of the systems admits a translation into the other holds good only with respect to formulas which are not based on the assumption that identity is a relation. Formulas which assume that, however, resist the type of translation which has been suggested by Hintikka. In fact at 5.534 of the Tractatus, Wittgenstein lists some such formulas, for example a = a, (x)(x = x), and (Ex)(x = a). In the Wittgensteinian framework these expressions (?) stand on par with expressions like 'the set of all objects which are not identical with themselves'. Yet, they are important because they indicate the precise sense in which Wittgenstein has given us an alternative notation, and also the exact sense in which identity is not a relation. If identity is a relation then expressions like these will have sense. But in so far as Wittgenstein succeeded in showing how the intended job of identity can be altogether dispensed with, and how what it is supposed to in virtue of being a relation can be accomplished otherwise, these expressions are devoid of sense. Sense, after all, is a constrained notion and is relative to a system of representation.

In the Wittgensteinian frame of things, = occurs as a meta-linguistic symbol; in 4.241 of the Tractatus Wittgenstein sums up his positive views on identity:

> "When I use two signs with one and the same referent, I express this by putting the sign '=' between them. So 'a = b' means that the sign 'b' can be substituted for the sign 'a'.
> (If I use an equation to introduce a new sign 'b', laying down that it shall serve as substitute for sign 'a' that is already known, then, like Russell, I write the equation - definition - in the form 'a = b' Def. A definition is a rule dealing with signs.)[33]
> Expressions like 'a = a', and those derived from them are neither elementary sentences, nor is there any other way in which they have sense."

This means that 'x = y' is another way of saying that 'x' and 'y' refer to one and the same object. Thus ' 'a = b' Def. ' defines the referential equivalence of two signs which occur in it, and thereby yields a rule permitting their mutual substitutivity. But, what after all is a definition? And what is a rule? If Wittgenstein gave too much attention to these issues, it is because he must have found Frege - should we say, even Frege - lumping together all sorts of ideas in his attempt to define identity so as to obtain a rule for substitution. That the identical are substitutable is never doubted by anyone; on the other hand there is almost universal agreement on that. The point of disagreement has always been the conditions of identity itself; most of the philosophical differences are rooted in a determination of these conditions. To bring this to the fore let us first note what constitutes a determination itself. Frege as well as Wittgenstein understood that term in the sense in which it is used by practicing mathematicians. For instance, it is normal in the mathematical practice to take, in the context of a plane, two lines defining a point, this is to say that they fix or determine a point. To this mathematical sense Frege gave an epistemological twist. Two lines present a point, he said, in a particular way. Such a mode of representing is what he called the sense of the term. There are different modes of presenting the same, and hence the need for criteria of identity. There might be reasons for preferring one particular definition, and a particular mode of representation, as a definition; for example, in the context of a plane, a point is an intersection of two lines. It is in this sense, definitional choices are ordinal, and senses are constrained; and in the context at hand what is stated in the definition is all that we know, and need to know, in order to work with points in a plane.

Frege had this on his mind when he thought that whatever might be the correct definition of identity, whatever might be the way in which it is presented, it must include a necessary component of its content the Leibnitzian Law as he understood it. This is the same as to say that according to Frege, all definitions of identity have an invariant component, namely $x = y \rightarrow (\phi x \rightarrow \phi y)$. Thus, in order to be able to tell whether x and y are mutually substitutable, variantly put in order to be able to tell whether x and y have the same truth-value, we either will have to know that a and b have the same referent, or else will have to know that they have the same sense. If identity of sense is what is required for substitution, then, as Frege came to believe that[34],

> "all identities would ... amount simply to this, that whatever is given to us in the same way is to be reckoned as the same ... a principle so dubious and so sterile as not to be worth stating."

This implies that Frege seems to have thought that identity of reference can be taken as a sufficient condition for substitutivity. It may be pointed out that then $x = y \rightarrow (\phi x \rightarrow \phi y)$ does not deserve to be called a definition in the strict sense of the term. While anticipating and agreeing on this, Frege writes:

> "I agree that Leibnitz's explication ... does not deserve to be called definition; my reasons, however, are different from Hussrel's. Since my definition is an identity, identity cannot be defined. This explanation of

> Leibnitz could be called an <u>axiom</u> that brings out the nature of the <u>relation</u> of identity; as such it is fundamentally important."[35]

Explication, definition, axiom, and relation! This awful unclarity or uncertainty about the nature of $x = y \rightarrow (\phi x \rightarrow \phi y)$ is the result of Frege's inability to set apart two notions, namely referential sameness and substitutivity. They belong to two different orders, the former figures in the meta-language and the latter belongs to the object language. Frege collapsed the difference between them and got into the muddle. This does not however mean that those two notions are unrelated. Far from it; as Wittgenstein noted in 4.421 of the <u>Tractatus,</u> = is indicative of substitutivity, substitutability, if one wants it that way. In other words, given the meta-linguistic rule '$x = y$', and the object language expression ϕx, we can move in the object language to ϕy [36]. The rule '$x = y$' formulated in the meta-language so to say legitimizes the object linguistic detachment of the consequent in $\phi x \rightarrow \phi y$. The four following remarks from the <u>Tractatus</u> amount to just that.

> 4.241 Can we understand two names without knowing whether they signify the same thing or two different things? - can we understand a sentence in which two names occur without knowing whether their referent is the same or different?
> 6.232 Frege says that two expressions have the same referent but different senses. But the essential point about an equation is that it is not necessary in order to show that the two expressions connected by the sign of equality have the same referent. Since this can be seen from two expressions themselves.
> 6.2322 <u>It is impossible to assert the identity of reference</u> of two expressions. For in order to be able to assert anything about their referent, I must know their referent, and I cannot know their referent without knowing whether what they refer is the same or different.[37]
> 6.2323 An equation merely marks the point of view from which I consider the question; it marks their equivalence in reference.

Note the expression 'marking a point of view'. I shall not raise the issue whether this is the same as Frege's mode of presentation, for what I am immediately interested in are two simpler issues, namely: What is Wittgenstein marking off here, and what is Wittgenstein's point of view? It seems that he is trying to draw our attention to the fact that the so-called identity sentences are no sentences at all.

> 5.473 The reason why 'Socrates is identical' means nothing is that there is no property called identical[38].

Similarly 'a and b are identical' is meaningless because there is no relation called identical. Now, does Wittgenstein hold that there is no relation of whatever level called identical, specially that there is no second order relation called identical! It appears from the passages cited above and other remarks that he is inclined to believe it to be so because, among other reasons, the second order sentence $x = y$ is not an indicative one. It is a normative sentence, and imperative sentence. Otherwise rules become descriptions of practices losing their normative force[39].

Although earlier I noted why identity sentences are meaningless, obviously decoding 5.5303 of the Tractatus, I now need to harp on that a little more so that we may have a better vision of the background of Wittgenstein's views on identity in 5.5302. In this attempt I shall pay attention to 2.02331 and comment on it. It was reported, by no less an authority on Wittgenstein's unpublished scripts than von Wright, that an early type-script of the Tractatus which was subsequently published as Proto-Tractatus, Wittgenstein prefaced in the margin of the passage which corresponds to 2.02331 of the Tractatus: "beilaufig gessprachen". I invoke this to suggest that the following comment on that passage may also be taken in a similar spirit. Our interpretation of Wittgenstein's text in a sense is isomorphic with the text; it is geared to indicate things which elude description in the strict sense of the term.

> 2.02331 Either a thing has properties that nothing else has, in which case we can immediately use a description to distinguish it from the others and refer to it; or, on the other hand, there are several things that have a whole set of their properties in common, in which case it is quite

> impossible to indicate one of them. For if there is nothing to distinguish a thing, I cannot distinguish it since otherwise it would be distinguished after all.

2.02331 and 2.0333, which is a comment on it, as the Hintikkas have impressed on us, provide the "metaphysical base" for Wittgenstein's treatment of identity; we may note that they provide a base for his treatment of descriptive phrases too. The broad context of these remarks is the issue about, given the fact that the logical form of an object is all that can be known about it a priori, it is only the internal properties of objects which can be expressed in language. 2.0333 reads as follows:

> "If two objects have the same logical form the only distinction between them, apart from their external properties is that they are different."

But what after all is the logical form of an object[40]? And whatever it might be, is it its internal property? What is an object's internal property anyway? No neat straight forward answers are forthcoming from the texts; and no amount of screwing those texts will yield incontestable answers. Yet, some such exegesis might give us some ideas which might receive a nodding approval, might give us something which might be conceded for argument sake, at least. In his letter of 16.1.1913 to Russell, Wittgenstein puts forward what he thought out as a solution to his as well as Russell's problems. This solution is based on the idea that different kinds of "entities" - individuals, properties, relations, etc. - be symbolized by different kinds of symbols which cannot possibly be substituted in another's place. This means that the logical form of entities of different types determine the way in which they can be combined with each other. Thus, first the replacement of objects is restricted to a type. Correspondingly substitution of names of objects is restricted to a kind. Now, whatever a determination of the logical form of an object might amount to, its name being not substitutable for the names of certain other kinds will be a part of that determination. (Note and then ignore how the conceptual batteries and conceptual confusions of the ancien régime linger on here.) Secondly, any two objects of a type cannot be replaced without altering the world; similarly any two names of a kind, or any two names of objects of a type, cannot be substituted for each other without tampering truth-value. The determination of the class of objects whose members can be mutually replaced without restrictions and constraints is a part of the determination of the form of the object. Such a determination determines its logical form. Analogously, a determination of the class of names of objects which can be substituted for each other without altering truth-value is a part of the determination of the logical form of a name - the semantic behaviour of a name, that is. What 2.0333 is asserting is that if two objects have the same logical form, then their numerical difference is ungrounded and an explanation of that difference cannot be located in their external properties. Similarly, if two names have the same logical form, the same semantic behaviour that is, then their difference cannot be located in anything else. In so far as this behaviour is restricted to referring - and according to Wittgenstein names only refer and only names refer, or to put is differently they do not express senses, as has been thought by Frege - we can restate it to mean that if two referring expressions refer to the same object, then they are two referring expressions referring to the same object, and that is the end of the matter. 'Scott' and 'the author of Waverley', for example. But this is not what Russell has done; he needlessly brought in his theory of descriptions, the iota operator etc.

> 5.452 The introduction of any new device into the symbolism of logic is necessarily a momentous event. In logic a new device should not be introduced in ... a completely innocent air.

Is it necessary to do what Russell did, then! Wittgenstein's answer is in the negative and it is based on his recognition that the ontological ungroundedness of the numerical distinctions of objects may have an analogue in language such that the introduction of the iota operator in the Russellian fashion is rather inutile. This should explain why, like Frege, he keeps definite descriptive phrases and names in the same basket. As he says in 3.31 of the Tractatus,

> I call any part of a sentence that characterizes its sense an expression (or a symbol).

Though this is perfectly in tune with the Fregean practice, Wittgenstein's handling of descriptive phrases is at variance with it. The semantic complexities in Frege are due to his belief that logical subjects are syntactically simple and primitive. Wittgenstein is reversing the Fregean position by holding that the logical subject may be syntactically complex but semantically simple. As I see it, this is what 3.31 amounts to. The criterion for deciding upon whether a given expression is the logical subject of a sentence in which that expression seems to appear is however not based on its simplicity or complexity; it is on the other hand solely dependent upon whether it determines the sense of the sentence in which it occurs or not. This to say that the logical subject of a sentence is to be isolated or identified in terms of its semantic decomposition. Take, for instance, the sentence

> The author of Waverley is a poet

and compare it with the sentence

> Scott is a poet

to note the sameness of the semantic behaviour of 'the author of Waverley' in the first sentence and 'Scott' in the second. In the former none of the first four words individually characterizes or determines the sense of that sentence. The four words together in their totality determine the sense of that sentence[41]. This is the case with that sentence as much as it is the case with the second sentence; the sense of the latter is not determined by the first five letters in their individual capacity. Those five letters in their totality, this is to say that 'Scott' in its individuality, determine its sense. So to say, as neither the author is a poet nor is Waverley a poet, 'author' and 'Waverley' cannot be taken as occurring - in the strict sense of the term - either in 'the author of Waverley' or in 'The author of Waverley is a poet' with their normal reference tagged to them. 'author' occurs in these expressions precisely in the sense in which 'Wit' and 'wig' occur in 'Wittgenstein' and 'Ludwig' respectively. This kind of an occurrence is not semantical; it is at most ocular. And it is no news to anyone that it is the mode of referring which is integral to the mode of representation. Here, the puzzlement seems to be not about syntactical complexities, but - as it emerges - of complex orthography. By being indifferent to orthography, Frege and Russell were tempted to over-determine the problem to the point of its eluding any plausible solution. If perchance we become aware of the obvious, we will have realized that what is true of names holds good with definite descriptive phrases too. And names, as Wittgenstein remarked in 3.26 of the Tractatus cannot be dissected further. They are syntactically and semantically impenetrable, just as objects are so ontologically (and logically). Moreover, a name, a descriptive phrase, an indexical expression, all are, within the Tractatus frame of things, essentially the same because

> 3.341 ... what is essential in a sentence is what all sentences that can express the same sense (i.e. represent the same fact) have in common[42].

Similarly, what is essential in a referring expression, to speak more generally in a symbol, is what all expressions referring to the same object have in common, or what all expressions serving the same semantic purpose have in common[43].

If the foregoing is plausible, either as an interpretation of the Wittgensteinian position or on its own, then it should follow that Russell's analysis of descriptive phrases does violence to syntax, specially in blurring the distinction between names, in fact referring expressions as such, and sentences[44]. That categorical blurr obtains in the Principia because within its framework, on the one hand (Ix)Px is definitionally equivalent to

$(Ex)\{Px.(y)(Py \rightarrow x = y)$, and on the other hand, say $Q\{(Ix)Px\}$ is a well-formed expression fully qualified to be admitted into the category of sentences. (Russell's analysis of descriptive phrases involves three things, namely a mapping, and a reduction, and an indifference to their difference!) We do not know whether Wittgenstein had this (too) on his mind when he wrote to Russell in his letter cited above that we need to devise a symbolism, or attend to orthographic matters, in such a way that it precludes the possibility of one category of expressions becoming substituends for another category of expressions. But Wittgenstein's objections to Russell on the count of his analysis of the descriptive phrases, and also his suggestions for an alternative symbolism fit in here. What then must have induced Russell to go the way he went to handle definite descriptions! Perhaps it was his acceptance of Frege's dual-function semantics and his conflation of the Fregean pair 'sense-reference' with the conceptual pair available to him from the philosophical tradition to which he belonged; we may recollect here that Mill was the first philosopher whom Russell studied. Anyhow, being a simpleton might be sufficient to fling 'Scott' onto Scott in order to refer to him, and it might require the cerebral churnings of a logician to invent an operator, or the scholarship of a literary historian, to spin an expression like 'the author of Waverley'. However, all this skill and labour are irrelevant to fling that descriptive phrase onto Scott so as to refer to him. Even here, only in so far as the flinging is concerned, the logician and the literary historian stand at par with the simpleton. In so far as the understanding of the mechanism of representation is concerned, we need not take into account how referring expressions are constructed; ignoring that aspect of the affair we can restrict our enquiry to the relationship between the expressions and what they represent; otherwise our enquiry will succumb to psychologism. This does not mean that psychology of language is inutile, and hence should not be indulged in. All that it suggests is that psychology of language should not be confused with "logic" of language. Both Frege and

Wittgenstein tried to come out of the influence of psychologism and subjectivism while seeking a solution to the problem with some Kantian ring around it, namely how does language come to successfully represent the world. Frege tried it by introducing truth-values as the referents of sentences, and Wittgenstein by introducing the concept of "darstellung" in contradistinction to the Kantian "vorstellung" (of course, not consciously, for there is no evidence that he ever read Kant)[45]. Their attempts were towards understanding the mechanism of linguistic representation independent of the user or "the subject" - towards paving the way for a vision of language, that is. All that is required to accomplish this is to pin down the relationship between sentences and the world taking into account only (i) the structure(s) in which referring expressions figure, and (ii) the relationship between the referring expressions and the world. The subject matter of our enquiry, thus, is the structure in which referring expressions figure. If these expressions have any inner structure it is outside the purview of our enquiry. Wittgenstein's point is that in so far as our enquiry is concerned, this is to say for the purpose of doing semantics, referring expressions do not have an inner structure; even if they have one, it cannot be analyzed. Nor can it be exhibited; just as even if objects have an inner structure it defies analysis because all that we can discern is restricted to structures in which they figure as unanalyzable simples. A referring expression thus is semantically simple though it may be complex otherwise. There can be names and descriptive phrases which when transcribed may run to the length of the current edition of the Britannica; the composition of a descriptive phrase has no role to play in its semantic functioning; modes of sign production do not determine the semantical.

This, in general, seems to be the way Wittgenstein thought about descriptive phrases. This does not, however, mean that he would suggest that all those expressions which Russell would treat as definite descriptions can be treated alike. That would lead to troubles - this time they would be brought in by orthographic similarities. What we need to keep in mind while handling descriptive phrases is the principle which we stated above, namely; isolate those parts of the sentence whose reference characterizes the sense of the sentence. Evidently this follows from Frege's constitution principle, which amounts to saying that the meaning of a compound is uniquely determined by the meanings of the components which constitute that compound. In such an attempt we may find that some expressions with the apparent form 'so and so' are such that their constituents, or expressions occurring in them, have their normal reference. Then, obviously, they themselves are not referring expressions, but are referring expressions containing referring expressions and standing in need of an analysis. As an alternative to Russell's predicative analysis of such phrases, Wittgenstein suggests a relational analysis[46], which has the virtue of restoring such constituents their normal reference which they lose in Russell's analysis. For example 'the present king of France', on Wittgenstein's suggestion, turns out to be an incomplete relational scheme where 'the king of' is the diadic relation, and 'France' the second relatum. This scheme will result in a sentence when the first relatum is supplied - say 'a (is a king of) France'[47]. Whether this is a better way of managing descriptive phrases is not of our concern here. What is of interest to us in the present context is that from what Wittgenstein is suggesting a simple message can be drawn out; it is that there is no blanket strategy to deal with all descriptive phrases. In each sentential context we will have to note how a descriptive phrase which figures in it as a part does in fact characterize its sense. If so, there is nothing logical about Russell's theory of descriptions. The reason for this is simple, namely that what is logical is not introduced for each case; it is to be given for a whole category and for once.

(II)

Wittgenstein could not set himself free from the grips of the devil (is it not the way he characterized identity!) with all the effort that went into the Tractatus. Though he completed the text[48] by the summer of 1918[49], he was still requesting Engleman to mail him a copy of Frege's (3) even on 25.5.1919. Of course we now have an idea of the vicissitudes his life has undergone after the first World War and before his return to Cambridge a decade later; yet it is, and for a long time it is likely to remain, only a rough and vague idea. This necessitates an attempt at a conjectural understanding of the development of his thought during the period in question. In any case his development like the development of Beethoven whom he admired is a continuous unfolding of a genius. Thus, it is unfortunate that Beethoven's as well as his life should have been compartmentalized into an early, a middle, and a later period[50]. Such a "Trinitarian" division of his life and thought is as useful in understanding and appreciating his contribution as the division of it into thirty parts. So in taking note of his post-Tractatus reflections on the problem of identity, I shall ignore all divisions. From our knowledge of Ramsey's visits to Wittgenstein in Austria, Wittgenstein's exposure to Brouwer's views on the nature of mathematics as well as mathematical activity, from our ability to surmise the way he could have reacted to Russell's and Ramsey's reaction to the Tractatus, and finally from what he wrote after he returned to Cambridge[51] we can conclude with some confidence, of course supported by minimal textual evidence, that he continued to have puzzlements on the count of both the Fregean axioms of identity, namely (i) $x = x$, and (ii) $x = y \rightarrow (\phi x \rightarrow \phi y)$. He neither altered[52] nor revised what he came to believe about these axioms during the Note Books - Tractatus period. But he was at loss to figure out how, given his views on them, the rest is to be adjusted. When identity as understood by

Frege and Russell is demystified and debunked, the extensionality thesis becomes problematic; and, with that, the logicist program which was thought out to provide foundations to mathematics will itself stand in need of foundations! (This does not mean that he subscribed to logicism; he was not an insider to this movement, but responded to the problems it raked as if they constituted, to use the current idiom, a research program.) It may be remembered that the logicists finally lost their hopes only in 1931. Thus the significance of what Wittgenstein was puzzled at, namely whether mathematics needs foundations at all! - and how this issue itself can be settled until after we come to know what after all mathematics was all about. He thus opened a mini-Pandora's philosophical box.

In so far as the issues with which I shall concern myself here are concerned, Wittgenstein's influence on Ramsey seems to be total; Ramsey's differences with Wittgenstein are only marginal and concern with fixing the nature of the inadequacies of the Principia[53]. Both of them found fault with Russell's definition of identity, the axiom of reducibility, the way Russell defines (away) classes, the theory of types, and the axiom of infinity. (One can wonder what else remains in that work!) As I will not be interested in the entire range of issues involved in these, I shall restrict my care and concern to the definition of identity and the issues related to it. To begin with, Ramsey echoed what Russell remarked in his introduction to the Tractatus. Considering 5.5302, in his review of that book for Mind, as an "important destructive criticism"[54] and "a rejection of" 13.01 of the Principia, Ramsey remarks that it

> "may have serious consequences in the theory of aggregates and cardinal number; it is, for example, hardly possible to say that two classes are only equal when there is a one-one relation whose domain is the one and the converse domain the other, unless such relations can be constructed by means of identity."

Yet, he thought - at least, he very soon came to see - that the treatment of = in the Principia has a "serious defect in it."[55] Recognizing that what was needed in the Principia is numerical identity, and thinking that what Russell intended the occurrence of = in the definiendum of 13.01 to capture us the philosophical notion of identity, Ramsey (too) suspected the validity of Russell's definition. Following Wittgenstein, he suspected the axiom of reducibility; he saw "no reason to suppose" that for every non-elementary or higher order function there exists an elementary function which could be true. He agreed with Wittgenstein in holding that

> "if it is true, this would be a happy incident and not a logical necessity for it is not a tautology."[56]

Believing that 13.01 "is to be rejected on other grounds" as well, Ramsey thought that the logical relationship between 13.01 and the axiom is what Wittgenstein thought it to be[57]. His point was that a rejection of the axiom entails the impossibility of only a "general proof" to the effect that two things sharing all their elementary functions also share all of their functions. Even after rejecting that axiom it would still be possible to prove that in any given specific case the result does hold. Ramsey in fact proved it for a particular case and concluded that:

> "hence rejecting the axiom of reducibility does not immediately lead to rejecting the definition of identity[59].

Thus it seems that Ramsey is supporting Wittgenstein, unknowingly though, because what Wittgenstein questioned was the logicality of the axiom, and thereby the logicality of the definition. What he thought was that what is logical is not to be introduced for each case, but only once for all cases of a kind. Precisely on that count Wittgenstein's argument is vindicated because what Ramsey has shown is that without the axiom, 13.01 does not have the generality which is required for anything for being admitted into the realm of the logical. But this point needs not be played upon further because Ramsey thought that rejecting 13.01 for that reason - this is to say that, on the count of the axiom - is only a minor matter.

Ramsey's major irritation with, or his "real objection" to, 13.01 comes to the fore in his paraphrasing of the parenthesis in 5.5302 of the Tractatus. He thinks that it

> "is a misrepresentation in that it does not define the meaning with which the symbol for identity is actually used. This can be easily seen in the following way: the definition makes it self-contradictory for two things to have all their elementary properties in common. Yet this is really perfectly possible, even if, it never happens. ... (S)ince this is logically possible, it is essential to have a symbolism which allows us to consider this possibility and not exclude it by definition."[59]

This means that Ramsey agrees with Wittgenstein in holding that the identity theory of the Principia is "a misrepresentation of mathematics"; but he is not willing to go along with Wittgenstein to replace = by a convention to the effect that disparate signs should have disparate referents. He is not inclined to do so because, as he noticed, that would lead to some inconvenience.

All that is given to Ramsey's difference with Wittgenstein may not be only what I have barely mentioned. There seems to be another and perhaps a more important reason for his reluctance to go along with Wittgenstein; and even that is based on a misunderstanding of what Wittgenstein had to say. when = is not available, x = y cannot be used as a propositional function in defining classes. What, then, is left to a mathematician! - obviously, only classes which can be defined by predicative functions. If so, quite a bit of mathematics, an important part for that matter, will have to be sacrificed. Analysis for instance. As it is, Brouwer and Weyl are out, from a different direction, to debunk it; thus there is that external danger. Now there is this Wittgensteinian "internal" subversion. Ramsey, prevailed over by the practicing mathematician in him, resolved not to give up the methods and the results of

Analysis in order to satisfy the "private prejudices" of Brouwer and Weyl, but confessed:

> "... I can neither accept the axiom of reducibility nor reject ordinary Analysis (i.e. Analysis as it was done in Cambridge at that time). I cannot believe in a theory[60] which presents us with no third possibility."

The situation, as Ramsey saw it, was thus desperate; and desperate situations need drastic remedies. To preserve mathematics in the classical sense of the term, logical infrastructure with the expressive power of the extended SQ at least is necessary. So, Ramsey came to see, the

> "only practicable way is to do it as radically and drastically as possible; to drop altogether the notion that Pa says about a what Pb says about b; to treat sentential functions like mathematical functions."[61]

This strategy extensionlizes the notion of a propositional function itself; and it does it <u>in toto</u>. Just as a function will not have the same value for two different arguments, ϕa and ϕb will not have the same value. This holds good for each ϕ; yet, it does not follow that a and b are one and the same. This itself is no solution to the problem. It at most provides, obviously in virtue of implying a clear rejection of the principle of identity of indiscernibles, a clean background for finding a way to the intended goal, which is to preserve the received corpus of mathematics while sticking to the extensionality thesis to the extent that mathematics "may be called a calculus of extension."

In view of the fact that all that would be available now constitutes just predicative functions, evidently

> "mathematics ... becomes hopeless because we cannot be sure (for e.g.) that there is any class defined by a predicate function whose number is two; for things may fall into triads which agree in some respect, in which case there would be in (the) system no unit classes and no two member classes."[62]

Of course, Ramsey is aware of the Wittgensteinian solution for overcoming this "cardinal" problem by an alternative interpretation of the quantifiers, and by working out an alternative symbolism. He concedes that identity is not necessary to specify the cardinality of a class defined by a predicative function. We can, for instance, formulate without introducing identity a sentence representing the fact that there are two books. But this is only a hand-to-mouth device which will not allow the mathematician to handle all the types of classes which he has been handling all these years. It restricts his activity to classes which can be defined either by enumeration or by a predicative function. Taking "class" to mean "the set of things of the same logical type", it is easy to see that there are classes which may not be definable either by enumeration or as extensions of predicates. They will have to be handled only through sentences about all classes or some classes. The richness of mathematics will be lost if we cannot talk about classes without requiring them to be extensions of actual relations or be in one-one relation in order to be infinite.

The only way in which all classes or all types of classes could be talked about is to have a logical device which in effect has the power of identity. To construct such a device Ramsey introduced the notion of a sentential function in extension, which "results from one-many relation in extension" between sentences and individual terms. It comes off "as a correlation", "practicable or impracticable", between individual terms and sentences. The former are the arguments of the function and the latter its values. Suppose in a given correlation we correlate 'Socrates' with 'Queen Ann is dead', and 'Plato' with 'Einstein is a great man'. Thus where ϕ is a function term of one argument,

ϕ (Socrates) = Queen Ann is dead,

and

ϕ (Plato) = Einstein is a great man.

Evidently such a correlation is arbitrary. Now let $\phi\hat{x}$ be the result of such a correlation, and let the suffix e be used to indicate that $\phi_e x$ is a function in extension such that we can now

> "talk of the totality of such functions as the range of the values of an apparent variable ϕ_e."

Thus $(\phi_e)(\phi_e x \equiv \phi_e y)$ amounts to saying the same as that, for each i, the sentence correlated to x and y are the same in the correlation C_i[63]. Now suppose that in some correlation C_i, x is correlated with a sentence S, and y is correlated with the sentence ~S, then $\phi_e x \equiv \phi_e y$ is self-contradictory. But if x and y receive the same sentence $(\phi_e)(\phi_e x \equiv \phi_e y)$ turns into a tautology. This induced Ramsey to remark:

> "it can suitably be taken as definition of $x = y$. $x = y$ is a function in extension of two variables. It is a tautology when x and y have the same value, contradiction when x and y have different values."[64]

This strategy of Ramsey segregates identity sentences from the realm of the contingent with some unpleasant consequences which subsequently formed the subject matter of Wittgenstein's severe criticism. This happens to be so with Ramsey's device because his acceptance firstly of universally quantified sentences as logical products, and secondly of the standard truth-functional logic. Where $\phi^{Ci}{}_e$ is $\phi(x)$ in a correlation C_i, $(\phi_e)(\phi_e x \equiv \phi_e y)$ will be $\phi^{C1}{}_e\, x \equiv \phi^{C1}{}_e\, y . \ldots . \phi^{Cm}{}_e\, x \equiv \phi^{Cm}{}_e\, y$. Thus even if one element of this conjunction is contradictory - and it will be so if, for some i, $\phi^{Ci}{}_e\, x \equiv \phi^{Ci}{}_e\, y$ will be $S \equiv {\sim}S$ - the conjunction obviously turns out to be a contradiction. And it will be tautologous only if for each i, $\phi^{Ci}{}_e\, x \equiv \phi^{Ci}{}_e\, y$ turns out to be $S \equiv S$; but this is impossible unless x itself is y. Because the only thing which is not arbitrary in this business of correlation is in correlating the same sentence to the same individual term; the product will turn out to be a tautology only when a single individual variable is involved. But this is too trivial a case to be taken into cognizance. There is however a way to come out of this impasse; and it is in allowing the arguments of the function term ϕ be the names of objects, and the correlation be between objects and sentences. But unfortunately this at the same time is the way to the point where we started, because the problem has been about sentences like

Cecero married Cateline ≡ Tully married Cateline

and

Socrates drank hemlock ≡ The husband of Xanthippe drank hemlock.

Suppose sentences like these turn up, as they would, in the Ramsey-product! How can we decide whether they are tautologies or contradictions? To be fair to Ramsey and his proposal let us first try to bring some precision to it, and then stretch it further. In that direction we may treat $x = y$ as a many-one function, and not as an unqualified function as Ramsey took it to be, with arguments in the set of individual terms and values in the set of objects such that '$= x$' is 'the object married to Cateline', and '$= y$' is 'the object married to Xanthippe'. Given some such, if not exactly that, additional adumbration a Ramsey product formula may be made to look like one which works in the intended direction, namely in securing each conjunct in the formula of the form $S \equiv S$. When this much is secured, a Ramsey formula turns out to be of the form:

$$S_1 \equiv S_1 . S_2 \equiv S_2 . \ldots$$

On the strength of the received definition of identity, which is usually but rather unjustifiably imputed to Leibnitz and which Ramsey is rejecting, saying that a and b are identical amounts to saying that

$$F_1a \equiv F_1b . F_2a \equiv F_2b . \ldots$$ [65]

The "isomorphism" between these two (it is possible to hold that one is either reading too much or unjustifiably assuming too much in relating these two in the required way) is indicative of the success of his proposal in regaining whatever is lost in sacrificing the received notion or definition of identity; and as he claims, only so we preserve mathematics from the "Bolshevik menace of Brouwer and Weyl."

Wittgenstein was no Bolshevik either in mathematics or otherwise, though he once or twice contemplated (crazily, I believe) of migrating to Russia during the regime of Stalin[66]. He was as critical of intuitionism as he was of logicism. He seems to have noticed that both these positions do violence to the nature of mathematical activity. But he did not reject Ramsey's proposal because it preserves the spirit of logicism, nor did he extend his views because they were nearer to anything that goes against intuitionism. He rejected Ramsey's proposal for several other reasons, one of which follows from the foregoing, and hence need not be elaborated any further. What emerged from our discussion of the Ramsey's proposal is that it turns out to be trivial if sentences are correlated with individual terms, and that if a non-trivial sense is to be imparted to it, identity will have to be treated as a function, which, as Wittgenstein has been insisting on, it clearly is not. Worse is that on either of these considerations the proposal turns out to be rather unnatural. This is why Wittgenstein came to bracket Ramsey's and Russell's definitions of identity; he bracketed them on the count that both of them use the strategy of "specification by extension", wondering whether this way of specifying is of any use. The identity sign defined by Ramsey, remarked Wittgenstein in his (10)

> "has no resemblance to the one we use to express a substitution rule."

In his view, expressed in stray remarks strewn all over the same text, the relationship between a function, its arguments, and its values should not be arbitrary.

> "The signs '`fa`', '`fb`', '`fc`' are no more function and argument than the words "Co(rn)", "Co(at)", "Co(al)", and "Co(lt)" are. Here it makes no difference. Here it makes no difference whether or not the "arguments" - 'rn', 'at', 'al', and 'lt' are used elsewhere as words."[67]

In the absence of a rule according to which arguments are correlated with values, Ramsey's proposal cannot but sound unusual and unnatural. An application of a rule correlating arguments and values takes at least three factors into account. This is what all those passages from Wittgenstein's writings on rule following when put together, twisted and squeezed amount to. These factors are, firstly the rule itself, secondly the way in which the arguments are identified, and finally the nature of the values. Ramsey's correlations are not rule governed; they are arbitrary. Further, Ramsey has not paid any attention either to the way of naming - specially, multiple naming - or to the way the objects could be identified; and correlating of sentences in the fashion of Ramsey, deciding the tautologyhood of the Ramsey product sentences etc., can be carried out only after issues relating to the identity of names, and the distinctness of objects are settled to the satisfaction of everyone. Without such a settlement it is impossible to assign truth-values to equivalences like

Cecero married Cataline ≡ Tully married Cateline.

In so far as Ramsey ignores or bypasses these issues, his definition[68] - he himself calls it so - becomes rather non-constructive. Then the very purpose of his offering it can be likened only to an attempt to consciously commit a mistake to cover another which was committed by habit. He wanted to defend classical Analysis against the intuitionist accusation that much of it is highly non-constructive; but his defense itself is open to the same accusation. Whatever securing foundations for mathematics might mean, this surely is not the way to success in that direction.

Wittgenstein was unhappy with Ramsey's proposal for another reason as well, namely on the count that it leads to what I may christen here as the model modal myth, or simply the Kripke-trick, according to which the identical are necessarily identical, and the distinct are necessarily distinct. Another modal myth is that if it is not necessary to do some thing, then it is necessary not to do it; following this precept, instead of explaining and exploding the Kripke-trick, we shall restrict our concern to Wittgenstein. His unhappiness concerning these modal invocations antedates his (10); Waismann[69] reports Wittgenstein to be saying that

> "Ramsey explains identity in this way:
> 'x = x' is a tautology.
> 'x = y' is a contradiction.
> i.e. the symbol '... = ____' is a tautology if there is the same letter on both sides."

He reports also that Wittgenstein complained

> "that on this approach = cannot be used to express the inter-substitutivity of two different signs."[70]

Foegelin[71] thinks that Wittgenstein's objection to Ramsey is based upon his misunderstanding of the latter's intentions and motives. According to him, Wittgenstein erroneously thought that Ramsey was attempting to provide an adequate definition of identity yielding all the rubrics required for substitution, but that this was not what Ramsey wanted to achieve. In holding this, Foegelin is following Ramsey, who in a letter to Schlick - which was passed on to Wittgenstein by the addressee - imputed such a misunderstanding to Wittgenstein. In the part that Foegelin quotes from the letter Ramsey writes:

> "I never really meant to suggest that $Q(x,y)$[72] was a way of saying that x and y were identical. I imagined that Wittgenstein had shown that it was impossible to say such a thing - I only proposed $Q(x,y)$ as a substitute for the symbol $x = y$, used in general propositions and in defining classes."

This much is clear from Ramsey's (1) where he is explicit that his "definition" is "only a logical device" which in effect is "equivalent to identity". So that should not be at issue. It might even be the case that Wittgenstein had misgivings about the nature and the purpose of the device. What really matters here is that Wittgenstein's objection that it is devoid of any mechanism for substitution has not been met. Why should this be so? Why did not Ramsey respond to this? It could be because Ramsey thought that it was not needed as he was contemplating a reformulation of the object language such that the resultant system does not require a rule for substitution! Wittgenstein in fact gave sufficient hints towards such a reconstruction of the syntax, and Ramsey, being a keen student of the Tractatus, was well aware of them. For instance, a language, whose primitive vocabulary does not contain equi-referring expressions such that no two terms refer to the same object, can do without a rule for mutual substitution of identical expressions. A move in that direction would have pleased Wittgenstein; but Ramsey did not attempt that either. He was thinking in terms of the Principia, and not of the kind of languages which the Tractatus is envisaging. The Principia kind of languages require a principle of substitution which Ramsey did not succeed in providing. That is what Wittgenstein seems to be unhappy about.

It might be possible to prune Ramsey's device further and overcome the deficiencies which Wittgenstein noticed in it; but it still remains to be settled whether an introduction of such a device is necessary at all! Ramsey's answer to this would have been in the affirmative; and he would have supported it with the argument that but for that both mathematics and its extensionality would be in danger. But is mathematics an extensional discourse at all! Foegelin thinks that even in the Tractatus Wittgenstein was skeptical about the extensional character of mathematics. In support of his reading he cites

> 6.031 The theory of classes is completely superfluous in mathematics.

He connects this remark with another to the effect that the generality in mathematics is not accidental generality. Foegelin's interpretation, however, seems to be rather rash and is not sufficiently supported either by Wittgenstein's writings or by his exegesis. But this much is certain, namely that Wittgenstein did not hold that the theory of classes is superfluous; and it seems that he held it for reasons which are different from the one that Foegelin mentions. Wittgenstein seems to have come to believe that in the context of the logicist program of defining the concept of number in terms of the concept of class one is indulging in something unnatural. This logicist reduction is something that Wittgenstein could not stomach even during the Tractatus times. According to the Tractatus view, number is a formal concept; and a formal concept is given as soon as an object falling under it is given. A formal concept does not stand in need of representation by functions or classes. If Frege and Russell tried to represent formal concepts in that fashion, it could be, as Wittgenstein seems to have thought, only due to a misunderstanding either of the nature of formal concepts, or of the way in which we do use them in our daily life. It is sufficient to "represent" them by variables; and that indeed is the way in which they are represented in actual mathematical practice, say. It is wrong to think, as Frege did think, that numbers get assigned to concepts under which objects fall; numbers on the other hand, according to Wittgenstein, get assigned to objects that fall under a number concept. This means that the use for which the logicists have put the class theory is indeed gratuitous[73]. Further, the definition of 'class' itself leaves much to be desired, logically that is. In fact it is informed by the peculiar way in which general sentences are understood, namely as products and sums. This may click in the case of accidental generality. But the generality in mathematics is not accidental; it, as we normally say, is conceptual, or as Wittgenstein would put it, is "extended generality".

> "The mathematical notation for infinite series like
> $1 = {}^1/_2 + {}^1/_4 + {}^1/_8 + \ldots$
> together with the dots is an example of … extended generality. A law is given and the terms that are written down serve as illustration. In this way instead of (x)Fx one might write "Fx & Fy & …"."[74]

Mark the expression "the law is given"; given for what? - for continuing the series! When the class of all sentences of the form Fx is given - and it is given when the law to generate the series Fx, Fy, … is given - then we can represent it as (x)Fx [75]. It does not follow that (x)Fx is the product of Fx, Fy, … What is required of a function is that its values be the results of "applications"[76]. If so, a function cannot be identified with its extension. (This opens the doors to intensionality, but that is a different matter.) Ramsey knows this for he notes in (1) that the purpose of introducing functions in extension with the intension of analyzing sentences about infinite extensions is defeated "when a function in extension is to be introduced by a list of definitions". Yet, what he did in his devising a device equivalent to identity amounts to taking the illustrations of a rule to produce a rule! In this sense the formula that Ramsey constructed to go as a counterpart of identity is not only devoid of extended generality but also is non-constructive. It has no internal relation to what it is intended to represent. So Wittgenstein thought:

> 'Ramsey's theory of identity makes the mistake that would be made by someone who said that you could use a painting as a mirror as well even if only for a single posture. If we say this we overlook that what is essential to a mirror is precisely that you can infer for it the picture of a body in front of it, whereas in the case of the painting you have to know that the postures tally before you can construe the picture as a mirror image."[77]

At this stage we may divert our attention to Wittgenstein's response to the other axiom of Frege-Russell theory of identity, namely $x = x$. His distrust of this axiom and its substitution instances remained unchanged throughout his life[78]. As earlier I cited his remarks about it in (5), let me take into account what he thought of it subsequently. His point was that though if at all we have to "choose" between $a = a$, and $a \neq a$, we "opt" for the former; it is a "perfectly stupid and useless" choice. This "basic law of logic" - Frege called it that - is cited in Part 1.216 of (3) as an example of a paradigm case of a useless sentence[79]. Concerning the other basic law, namely

$$x = y \rightarrow (\phi x \rightarrow \phi y),$$

Wittgenstein suggested a replacement of it by a rule in the meta-language. This precisely is his solution concerning this law too. Foegelin was the first to recognize this, at least as transparently as he does, and so far as Anglo-American studies of Wittgenstein are concerned. I shall first state Foegelin's clarification of this point, and then use it for a purpose not intended by him, namely to make intelligible the parenthesis of 3.203 of the Tractatus. Foegelin arrived at that understanding on the basis of the following passage which can be found in (4) as well as (10).[81]

> "The way of making the two expressions similar is somewhat analogous to that which one sometimes adopts in mathematics, say in the proof that the sum of three angles of a triangle is 180°.

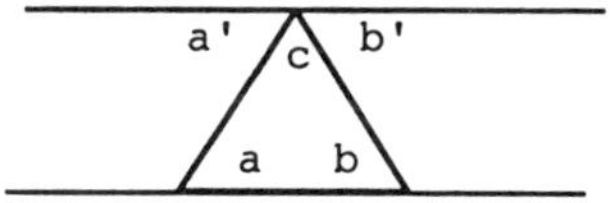

> We say $a = a'$, $b = b'$, and $c = c'$. (The first two equations are of an entirely different kind from the third.)"

The reason for this, as Foegelin sees it, is that the first two equalities are substantial elements in the proof; but the third, though important for the proof, can as well be dispensed with in favour of a convention, say by introducing a rule to the effect that two angles are equal if their non-shared components are equal. Such a rule entitles us to replace c = c as it occurs in the proof by a statement that c is a shared component of two angles.

To consider the parenthesis of 3.203, this is to say that the remark 'A' is the same sign as 'A' , if it is the case that we just noted holds good with self-identity, then what Wittgenstein has to say in this remark can be seen as the result of his attempt to encapsulate such a convention. Black (1) thought that 5.53 - 5.534 of the Tractatus are intended to answer the question: what shows that in A = A, 'A' is the same sign as 'A'. But this seems off the mark as 4.234 and 5.53 are geared to show why we do not, in fact why we cannot, ask that question.

> 4.243 Can we understand two names without knowing whether they signify the same thing or two different things? - can we understand a sentence in which two names occur without knowing whether their referents are the same or different? ....
> Expressions like 'a = a', and those derived from them, are neither elementary sentences nor is there any other way in which they have sense. ...
> 5.53 Identity of object I express by identity of sign, and not by using a sign of identity. Difference of objects I express by difference of signs.

The parenthetical remark in 4.243 and the remark about self-identity in 5.534, namely

> "and now we see that in a correct conceptual notation pseudo-sentences like 'a=a', and 'a=b & b=c .→. a=c', '(x)(x=x)', '(Ex)(x=a)' etc., cannot even be written."

might have occasioned Black to think that these remarks are intended to throw light on 3.203. But with a little more reflection it will be possible to notice that the parenthesis, on the other hand, is about what Wittgenstein has to say about such sentences. This should be evident from the example he himself gave, this is to say that from the fact that the first letter of the Roman alphabet occurs within quotes; it is not used, but only mentioned. That 'A' is the same sign as 'A' is not a statement about objects (that form the substance of the world); it is about signs. How objects are referred to by nomial signs is specified in the principal part of 3.203; it states that a name refers to an object, and that the object so referred is its referent (in contradistinction to Frege's belief that on occasions names refer to senses, and in one sense it is the concepts that refer to objects!) The parenthesis is intended to convey that in the concept-script a nomial expression will refer to the same object in all of its occurrences (and on all occasions!) This is to say that it is the name-type that has the specified semantic relationship with the object, and not the name-token. Variantly put, the constituents of a sentence are name-signs and not name-inscriptions. It is because of this that 'a = a' is useless, and so even if it is understood to be an assertion to the effect that the first occurrence of 'a' refers to the same object as the second occurrence of 'a'. Wittgenstein is differentiating between a sign and its inscriptions. Otherwise, on his own argument, the parenthesis either says nothing at all, or asserts an explicit falsehood. It cannot be taken to mean either that a sign is the same sign as itself, or that two signs are the same sign.

But after all why that remark, even if we understood in that way, is necessary? Why is it needed at all? What legitimizes its being in order, and what justifies its being in tune with the rest of the Tractatus, can be elicited when we note that the central preoccupation of Wittgenstein in that work is towards a clarification of the notion of representation in the concept-script. It is obvious that the very possibility of representation requires that the nomial expressions of the concept-script have referents, that their referents be constant, and that their referring be unique. (In fact these are the very presuppositions of SQ which is a theory of reference or representation.) The parenthetical remark in question concerns the second requirement. A convention to that effect secures the concept-script what is required to successfully represent the world. Thus though it is possible to hold that what Pa says about a may not be the same as what Pb says about b, it is impossible to hold that the a talked about in Pa is different from the a talked about in Qa. The bamboozling of Socrates might be, as it indeed is, different from the bamboozling of Plato, but the snub-nosed Socrates and the hen-pecked Socrates will have to be the same Socrates - the same object, the same individual. (If, per chance, they are not the same, then it becomes necessary to devise a strategy to successfully telescope them, to make them coincide; otherwise any further move in the direction of representing the world would be impossible.) But for this, objects cannot form the substance of the world; and that objects are the substance of the world is one of the central theses of the Tractatus[82]. Without this the constancy of the world would be lost, and the world becomes Heraclitean; correspondingly truth-values of sentences turn indefinite!

Frege showed how the existence of an object can only be exhibited by means of language through the use of its name; it is not, as he impressed on us - successfully, I suppose - something that can be asserted within language. Wittgenstein is showing us how the case is the same with identity. The identity of an object by the use of its name! So to say, in naming an object - he is impressing on us - we are "naming its identity" too. Also central to the thesis of the Tractatus is that the structure of the world is antecedent or prior to the structure of the concept-script. In the beginning there were objects, and not names; and thus neither language nor logic can claim precedence. This is the Wittgensteinian creation myth. We design logic and language to reproduce the world. Of course, there is no logical (or any other kind of) guarantee that we can succeed in capturing the world in the logical network which we have designed. But if we are successful, it is only a favour granted by fate! Anyhow the point worth noting is that the concept-script designed with that purpose needs not include identity. But not including identity in the concept-script, as Russell noted in his introduction to the Tractatus,

> "removes one method of speaking of the totality of things, and it will be found that any other method that may be suggested is equally fallacious. So, at least, Wittgenstein contends, and I think, rightly."

Thus we are constrained or condemned to speak sub species; and that is an important message of the Tractatus. Yet, if we find violations and transgressions of this transcendental limit, it is because, as Wittgenstein realized,

> "(a) philosopher says 'look at things like this' - but in the first place this does not ensure that people will look at things like that, and in the second place his admonition may come too late; it's possible moreover, that such an admonition can achieve nothing in any case and that the impetus for such a change ... has to originate somewhere else entirely."[83]

(III)

Coda:

Wittgenstein and the Principles of the identity of Indiscernibles:

Wittgenstein's understanding of the principle, and the way he uses it in his philosophy, for all these years, has been a dark spot in Wittgenstein-studies. It is true that he remarked in the Tractatus,

> 5.5302 Russell's (Principia) definition of ' = ' is inadequate, because according to it we cannot say that two objects have all their properties in common. (Even if this sentence is never correct, it still has sense.)

But it is doubtful whether he meant by this remark that Russell in his introduction to the Tractatus took him to be intending to say. Russell rightly thought that Wittgenstein subjected his "conception of identity ... to a destructive criticism from which there is no escape". However, the reasons for which Wittgenstein rejected Russell's definition are different from what Russell thought them to be. Russell believed that

> "(t)he definition of identity by means of the identity of indiscernibles is rejected (by Wittgenstein i.e.) because the identity of indiscernibles ... (is said) to be not a logically necessary principle. According to this principle x is identical with y if every property of x is a property of y, but it would, after all, be logically possible for two things to have exactly the same properties."

Wittgenstein's objection was not based on an argument to the effect that the principle is not logically necessary and that hence 13.01 of the Principia stands without logical force. This in fact seems to be one of the reasons for his being unhappy with Russell's introduction to the Tractatus which he thought was a clear misinterpretation of what he had to say on this as well as on many other issues.

Russell's version of the matter somehow has come to be endorsed by every interpreter of the Tractatus; this is the case in the English speaking world at least. (One of the reasons for this could be that it fits in with the basic tenets of the British empirical tradition; note the number of the twentieth century English speaking philosophers who thought that the principle is only a contingent truth - Armstrong, Black, ..., Moore, ..., Russell, ...) And Black's (1) reinforced this misunderstanding. In their defense, and also in defense of Russell, those who subscribe to this view cite 5.5302 of the Tractatus and foist on it an interpretation to yield the evidence they want to sustain that Wittgenstein is saying that because 'two objects have all their properties in common' is a sentence with a sense, and because for him a sentence with a sense is empirical (a doubtful assumption, for it drags Wittgenstein into the empiricist tradition) he is taking the principle as a matter of fact, and thus not a necessary principle. (In the last inferential jump, there is a bit of Hume, but I let all this go as questioning this is not really relevant to any purpose here.) But what is the sense in which the sentence in question has a sense! In the text we find it italicized. Could it be that Wittgenstein wants to indicate that he is using it in this context in a non-technical or in the commonsensical sense, unlike in the other places in the text where it occurs as technical expression! It appears that he wants to contrast its use in this occurrence from its use in other occurrences. Thus whereas in its technical uses we find it inscribed in regular type, this particular use resorts to inscribing it in italics. Wittgenstein seems to be indicating that he is saying what he is saying here with a "so to say" omitted. This should be evident from the very next passage which obviously is a comment on this remark.

> 5.5303 Roughly speaking to say of two things that they are identical is nonsense, and to say of one thing that it is identical with itself is to say nothing at all.[84]

What Wittgenstein wants to pack into this passage is that if the definition of numerical identity in terms of the principle is adequate, then to say that two things are identical (this is to say that they are one and the same) is the same as to say that they have all their properties in common. But to say of two things that they are identical in this sense of the term does not amount to saying something which is devoid of sense in the technical sense of the term; it is, on the other hand, equivalent to saying something senseless in the colloquial sense of the term. It amounts to saying something unsinning rather than something sinnlos; thus what he has to say is much worse than what has been hitherto thought. In so far as numerical identity and qualitative identity get conflated, as they evidently do in Russell's definition, it does not really make sense to say that two things have all their properties in common, because if the definition is correct, then there are not two things but just one; and saying that thus at most amounts to saying that a thing has the properties it has, no more or no less. On the other hand, if numerical identity and qualitative identity are kept apart, then it, of course, makes sense to talk of two things having all of their properties in common, for instance two photons, even though it may not always be right to talk that way, for example, as in the case of two peacocks surely. This is so because, in our day to day representation of the world, we seem to assume the dissimilarity of the distinct. Moreover - and this is important to note - when numerical identity and qualitative identity are separated, what acquires contingent character is not the principle, but Russell's definition itself. His definition, then, will have the status of say a theoretical postulate, and in any case will stand without the required logical necessity in the strict sense of the term. All this, then, means that what Russell thought to be Wittgenstein's objection to the principle is indeed his objection to Russell's definition itself. Badly missing this, Russell wrongly believed that Wittgenstein was rejecting the principle. The import of this, as can be surmised without any difficulty, is that Russell's reading of Wittgenstein's intention does not match with the content of the Tractatus. There is, on the other hand, sufficient internal evidence in that text to sustain the view that his conceptual net entails the principle. This may be entertaining too much; so let me make a weaker assertion. The principle is not inconsistent with the conceptual framework of the Tractatus. For instance, according to Wittgenstein, no two objects configure in the same way; nor do they configure in the same set of facts, Thus no two objects can have all of their "properties" in common, where 'property' is understood in the Wittgensteinian way. It may be recalled that his notion of an object is broad enough to include ordinary objects as well as their properties in the ordinary sense of the term. This is not incompatible with the principle or its equivalent, namely the principle of the dissimilarity of the distinct.

From the Wittgensteinian point of view the issues whether the principle is true or false, whether it is necessarily true, or is so only contingently - this is to say that the issues which preoccupied every leading British philosopher from Moore to Wiggins - are simply irrelevant. His concern was not with the principle itself. What he was interested in was the principle of substitution, a principle to determine referential equivalence classes of nomial expressions; the principle of the identity of indiscernibles was needlessly confused with the principle of substitution. (May it was the outcome of an indecision about the nature of identity - whether it obtains between objects or their names.) His grouse against Frege and Russell has been that they miscall the substitution principle the identity principle. On that count, as we shall see soon, he has a point. The former is a logical principle, and the latter epistemic. This misnaming leads to several confusions, one of which is: when the logical principle of substitution fails in the so-called intensional contexts, doubts are cast on the principle of identity. It has also been the case that when such failures are encountered, the ontological principle that no two objects have exactly the same properties was suspected - for instance by Russell himself. This ontic difference is ungrounded and is independent of discernibility. When it is noticed that an object "has a property that nothing else has", and a description is used to isolate it, its uniqueness is not established; and it is not established because if it were not unique, or as Wittgenstein would say "if there is nothing to distinguish a thing," it could not have been distinguished at all, as he remarks in 2.02331. But for this belief, he could not have demanded (in 3.22, 3.324 and 3.325) a language which captures such uniqueness, one as one, two as two, and n as n. This is to say that he is demanding a language whose form mirrors the logical form of objects, a language which shows that objects are numerically distinct whether we can distinguish them or not. That the language that is given to us, the ordinary language, is one such, and that Wittgenstein thought that it was one such, are matters which need not concern us here.

There is yet another, and perhaps more interesting as well as important, reason for entertaining that Wittgenstein did not reject the principle of the identity of indiscernibles. Such a rejection necessitates at some point or the other an acceptance of the <u>substrata</u> theory in one form or the other. That

Wittgenstein did not subscribe to that theory in form should follow from the following: (i) in the framework of the Tractatus objects themselves are the substance of the world. These objects are not composites, least of all composites constituted of universals and substrata - and neither of these alone form the substance of the world; (ii) the Tractatus notion of an object envelops what others usually call properties, too; of course, it is possible to argue that Wittgenstein did not subscribe to the bundle theory either, and hence it cannot be said that he subscribed to the principle of the identity of indiscernibles. The basic assumption of this argument can only be that the principle entails the bundle theory, this is to say that to subscribe to this principle means the same as holding that a set of properties uniquely individuates an object. In fact, it is Meinongianism, for which Wittgenstein never had any respect, and for which Russell had some fascination for some time to the extent of being considered by Meinong that he was his best representative in the English world. Meinongianism is compatible with the bundle theory and is incompatible with the principle of the identity of indiscernibles. To note how this obtains let a "proper predicate" be a predicate which does not contain either logical constants or singular terms. Meinong holds that for any set of proper predicates there exists an object having exactly those properties which correspond to those predicates. This implies that the power set of the set of properties can be one-one mapped onto the set of individuals. Then by Cantor's theorem the set of individuals cannot be one-one mapped onto the set of proper properties; this is to say that there are more individuals than properties. But this goes against the principle of the identity of indiscernibles, because it does not permit individuals to share properties; and if there are more individuals than properties either they will have to share some of them, or else some individuals will have to go unemployed and bare!

However, Wittgenstein did not in any of his writings explicitly subscribe to the principle, and he did not do this, I trust, because he need not do it. In the

Tractatus vision of things, neither does a set of properties individuate an object, nor does a set of objects all the members of which share a property individuate that property. Had Wittgenstein subscribed to any of these, the simplicity of the Tractatus system of ontology would have been adversely affected because then the logical and the ontological homogeneity of objects and properties, in the ordinary sense of these terms, would have been lost. His concept of objects, to repeat, envelops both of these. He held this view of objects from the period of the Note Books (see for instance the entry for 16.5.1915) to the time of the notes he dictated to Desmond Lee twenty years later. Thus, in an important sense, the principle is irrelevant to what he has to say about objects. What he was interested in was a principle of substitution, and on this issue he is one with Leibnitz who, in his Fourth Letter to Clarke, remarked:

> "to suppose two things indiscernible, is to suppose the same thing under two names."

Wittgenstein seems to have taken this idea of Leibnitz and assimilated it to the idea of ungroundedness of numerical distinctness which, it seems, was also borrowed from Leibnitz, either directly or indirectly[85]. To buttress this conjecture with some supporting evidence let me first consider the notion of numerical distinctness in Leibnitz.

What Leibnitz holds is much stronger than what Russell makes it to be, and there is more to Leibnitz's thesis than it appears to Russell's eyes. But then what Russell holds it to be is sufficient for, and suits, his purpose which is to be able to take the principle of the identity of indiscernibles as the obverse of Leibnitz's principle of numerical difference. Russell believes that the principle

> "asserts that there are not in nature two indiscernible real absolute things ... no two substances are completely similar or differ solo numero."[86]

But this does not close the possibility of two objects having a common property and thereby collapsing into one, when they have all their properties in common; on the other hand it leaves open such a possibility. What Leibnitz is adamant about is different from what Russell is saying it to be. Leibnitz's argument is to the effect that no two substances can ever have a property in common, because all properties are individual; Leibnitzian properties are Aristotelian too. He writes[87]:

> "Two subjects cannot have in the strict sense of the term the same individual property, it being impossible that the same individual property should be in two subjects."

For him all genuine properties are individual, and the rest are "fictions". If this were not so, there would be common properties; and the existence of common properties paves the way for external causation which, for Leibnitz, is more heinous than the cardinal sins. Thus even if two monads have the same contents, which obviously are perceptions, they need not be of the same clarity and distinctness. In this way numerical difference between two individual substances entails dissimilarity among individual properties of these substances. (Juxtapose 5.5303 of the Tractatus which is cited above!) Given this, if the principle of the identity of indiscernibles is the obverse of this principle, then it should assert not what Russell thought that it asserts, but something else; and this holds good for what Russell says that it asserts, in his own book on Leibnitz as well his tangential remarks on Leibnitz. For instance, in his introduction to the Tractatus he understand it to be saying:

> "if every property of x is a property of y, then x is identical with y."

But as it emerged what it really amounts to is saying that

> IF EVERY PROPERTY OF X IS A PROPERTY OF Y, AND FOR EVERY PROPERTY P OF X, P AS A PROPERTY OF X IS THE

> SAME AS THE PROPERTY P OF Y, THEN AND ONLY THEN X AND Y ARE IDENTICAL.[88]

That being an honest to God statement of Leibnitz's principle of the identity of indiscernibles, it should follow that neither Wittgenstein rejected the principle (though Russell thought that he did), nor Russell accepted it (though he thought that he did.) The entire debate concerning the principle has been one of those comedies of misgivings and rash interpretations which abound in philosophy.

Nevertheless, I find that Wittgenstein used the principle at least once in the Tractatus, namely at 4.1241, and as one of the steps in an argument. Not staking much on my reading of this passage in this way, it might for all that I do not know lead to an alternative and perhaps more plausible interpretation, let me try to draw attention to the way he uses a weakened form of the principle. He is raising here the issue concerning the distinguishsability of forms, and is using the principle understanding it a bit differently.

> 4.1241 It is impossible to distinguish forms from one another by saying that one has this property and another that property; for this presupposes that it makes sense to ascribe either property to either form.

Suppose, for argument sake, that ascribing properties to forms is a meaningful activity; then, Wittgenstein may be taken to be suggesting that the difference in properties may go towards distinguishing forms, because qualitative difference is indicative of numerical difference. But it is not anyhow a meaningful activity; so forms after all cannot be distinguished in this way. Anyhow, here he is not even suggesting that qualitative difference is numerical difference, which is what the principle, on Russell's understanding of it, amounts to.

(IV)

An Apology in Lieu of Summing Up:

Walter Benjamin once wondered why a whole book could not be written just in quoted passages; and as one who wondered why a whole philosophical treatise could not be written entirely in interrogatives, Wittgenstein would have thought Benjamin's idea brilliant. But bereft of the literary skills of Benjamin, and also obviously the philosophical talent of Wittgenstein, I had to be content with using the conceptual piecemeal, but nonetheless important, work done by White, Williams, Foegelin, and others. Using their work so to say as pillars, I tried to sketch a huge structure. Of course, I pruned their arguments and added to their work where I thought it was in order, or was warranted, and dropped some of their results which I felt could not be sustained in the light of a closer study of the texts. If I have anything to claim for myself, it is the innumerable interlinks, interpolations, and the introits which I provided to connect their independent and unrelated points in order to project and present a new whole. I have imposed an order on the otherwise chaotic material; the result, I trust, should give - to purloin a limpid phrase of Wittgenstein - "a vivid image of a huge building" which he contemplated. All this were as if Michelangelo found four pieces of marble in four different quarries sculptured so to say by geological convulsions into a hand, a foot, a nose, and an ear, and then set himself to keep them in their proper place filling in the gaps to given them the shape in which we now find his "David"! Anyhow, it is "David" that matters, and Michelangelo would have credited it to Nature and claimed nothing for himself.

Notes

[1] 'Abhandlung', 'Bemerkungen", 'Untersuchungen', ... are the terms which figure in the title that he gave to his works.

[2] To be precise three books and one paper of each of these authors, namely (1) - (4) of Frege, and (1) - (4) of Russell. Juxtapose the above extract from his letter with the following remark from Frege's (1), Sec.1:

> "No matter how indeterminate the meaning of a letter, we must insist that throughout a given context the letter retain the Bedeutung once given to it."

It may not be out of place to note that even in his (2), pp.63-69, Russell was puzzled by =; does it relate names or objects? He does not seem to have a clear idea. This vagueness persists in 13.01 of (4).

[3] As C.F.J. Williams comments, Frege answers (a) negatively, believes that he is able to circumvent the reasons for answering it negatively, and really thinks that the answer is in the affirmative. Williams sketches how "the distinctions between sense and reference are introduced with these credentials."

[4] See the entries dated the 26th and the 27th of October 1914 in (5).

[5] To preclude confusions, I shall use 'Begriffsschrift' to refer to Frege's (1), and concept-script in the generic sense.

[6] Names, for Frege, are what we now a days call referring expressions; thus 'Gottleib Frege', 'the king of France', 'my daughter's favourite joke', 'Pegasus', ... all are names. He holds that some names are complete, and others incomplete. For Russell, and subsequently Quine, none of them is really a name. Further, Frege seems to be fond of conceptual pairs; note how many pairs he introduces - sense-reference, concept-object, function-argument, ... He introduces a new pair whenever he encounters a problem. This creates the impression that his semantic explanations are ad hoc. One may even wonder whether he ever thought of a well-connected semantic theory.

[7] The gap between intention and practice is not idiosyncratic of Frege's (1); it is pervasive throughout his corpus as a whole. In his 1884 (2) he suggests adopting Leibnitz's law of identity of indiscernibles as a definition of identity; but in his 1893 (3) the second order $x = y \rightarrow (\phi)(\phi x \rightarrow \phi y)$ is said to be the chief law giving identity. In between he argued that because a definition itself is an identity, identity cannot be defined!

[8] Quine, in Word and Object, p.116 ff, has a rider to this; he remarks: "What makes identity a relation, and ' = ' a relative term is that ' = ' goes between "distinct occurrences of singular terms, same or distinct", "and not that it relates distinct objects." But on the count of referential interpretations and non-empty domains the problem remains where it was.

[9] Leibnitz's law as understood by Frege may yield all the properties of identity; but as Quine showed in From A Logical Point Of View, neither a relation R's satisfying it, nor its satisfying the schema $\phi y \rightarrow (Ex)(xRy \,.\, \phi x)$ completely telescopes R, within the framework of SQ, into what we normally intend or understand as identity. This should not, however, effect

the possibility of the Fregean claim that we need at least the schema (1) to mark off what is peculiar to identity.

[10] This is to say that by substituting an improper predicate for a proper predicate, where by 'improper predicate' we mean 'predicate in which a singular term occurs with its normal reference'.

[11] Russell (3).

[12] All that is dear to Frege's heart is that "(i)f we are to use the symbol a to signify an object, we must have a criterion for deciding in all cases whether b is the same as a...". (2), Sec.62. We should keep in mind that here a and b are taken as names of abstract objects like numbers for instance, and not as names of any sort of objects, say Scott, Pegasus, or an electron.

[13] There is enough textual evidence in support of what White thinks; but then in the case of Wittgenstein nothing can be settled by citing texts because in reading him one may have to read into him. The Tractatus specially is in the spirit of classical philosophical works in Sanskrit where only the outcome of a long train of thought is recorded without any indication of how that has been arrived at. Incidentally, Ramanujan who reached Cambridge in the year Wittgenstein left, entered in his Note Books that he left too many theorems without giving proofs as much as Wittgenstein recorded in his Note Books too many philosophical theses without giving any arguments. If providing proofs of those theorems is doing good mathematics, I do not find any reason why supplying arguments in support of Wittgenstein's remarks should not be doing good philosophy!

[14] Was Wittgenstein self-conscious about it when he remarked in (5) on 15.11.1914 that a "single insight at the start is worth more than ever so many in the middle"!

[15] Obviously, by this expression Wittgenstein would not have meant what we now a days call individual variables because he called these by this very name.

[16] Here is an echo of Frege and is indicative of the fact that he was thinking in the context of Frege; Frege's (4) seems to be on the forefront of Russell's mind.

[17] (4), pp.63-64; Wittgenstein was familiar with this work. It was reported that he read it during his Manchester period itself. The reference to an illegitimate process in the passage quoted here might have been occasioned by a recollection of the Bradeleyan regress.

[18] Vol., p.168.

[19] We may ignore here how Russell shifts from predicate-talk to property-talk within the pace of two successive sentences; anyhow that is possible for him, for he subscribes to one-one correspondence between predicates and properties.

[20] As if this were not sufficient Russell (see his (7)) imputes that to Wittgenstein; and not being content with it, he ascribes to Wittgenstein an unawareness of his own assumptions. He writes on p.115:

> "At one point I accepted (Wittgenstein's criticism at 5.5302 and 5.530 of the Tractatus) ... but I soon came to the conclusion that it made mathematical impossible and, in fact, that Wittgenstein's criticism is invalid ... Wittgenstein's

position assumes that diversity is an indefinable relation, although I do not think that he knew he was making this assumption. But if he is not making it, I do not see on what grounds he can say, as he does, that it is significant to say that two objects have all their properties in common. If, however, diversity is admitted, then if a and b are two, a has a property that b has not, namely that of being diverse from b. I think, therefore, that Wittgenstein's contention as to identity is mistaken. And, if so, it invalidates a large part of his system."

[21] (1), p.58.

[22] For Leibnitz as well as for Wittgenstein the principle of identity of indiscernibles turns out to be a generalization of the axiom.

[23] (5), p.192. Note the exclusive predicate-talk; here only predicative definitions are involved. Wittgenstein happened to see this book which Russell mailed to him after he completed the Tractatus. But he already came to his view on self-identity in (5), namely that $x = x$ is not even a proper sentence.

[24] The same thing can be said in relation to the theory of types. In January 1913 he wrote to Russell that "every theory of types must be rendered superfluous by a proper symbolism." (5), p.213.

[25] (5), p.125.

[26] Given his commitment to realism, and the robust sense of reality which he preached, it is natural that Russell should have construed all relations ontologically as relational properties and created a bizarre world which must have frightened Wittgenstein.

[27] p.23; emphasis added.

[28] Muehlmann thus holds that Russell is wrong in thinking that Wittgenstein rejected his definition because the fact of the matter is that he rejected not the definition but what Russell says about the definition. But there is another sense in which Wittgenstein can be said to have rejected the definition itself. Such a rejection follows from his refusal to accept Russell's view that $x = y$ is an abbreviation of $(\phi)(\phi x \rightarrow \phi y)$, and from his belief that = be reserved for meta-linguistic uses.

[29] When quantification theory is restructured along these lines, the theory of classes as developed by Russell cannot be expressed in the framework of the Principia unless it too is restructured. Anticipating that, Wittgenstein pays attention to that restructuring also, of course while holding that the class theory itself is superfluous. In the entry for 21.10.1914 in (5) he wondered: "isn't Russellian definition of the null nonsensical? Can we speak of a class $\hat{x}(x \neq x)$ at all? - can we speak of a class $\hat{x}(x = x)$ either? For is $x \neq x$ or $x = x$ a function of x? - must not 0 be defined by means of the hypothesis $(E\phi) : (x) \sim \phi x$? And something analogous would hold of all other numbers. Now this throws light on the whole question about the existence of numbers of things.

$0 = \hat{a}\{(E\phi) : (x) \sim \phi x \;\&\; a = \hat{u}(\phi u)\}$ def.

$1 = \hat{a}\{(E\phi) : \phi x : \phi y \;\&\; \phi z \rightarrow y = z : a = \hat{u}(\phi u)\}$ def.

It may be noted that the sign of equality in the curly brackets can be dispensed with in terms of

$$\bigwedge$$
$$0 = \hat{u}\,(\phi u)\,\{\,(x)\sim\phi x\,\}.”$$

which is to be read as: the class of all classes of elements u such that ϕu such that nothing is ϕ. The sign of identity as it occurs in the definition of 0 and 1 above is within the scope of the quantifiers. Earlier it was remarked that identity for Wittgenstein is not a relation. Here he is emphatic about its not being a function either. For him it is not even something which issues in tautologies; see the entry for 2.11.1914 in (5).

In the way the Tractatus envisions ontology, there are no negative properties. It seems that too many British philosophers are growing fond of these properties. Russell is not alone in thinking that “not being identical to a” is a property. There are several currently famous philosophers from the island cultivating such macabre creatures ; Noonan (see his Objects and Identity, p.47) thinks that “lacking a property P is itself a property”, and Wiggins (see his Substance and Identity, p.67) is one up in holding that Aristotle “has the property of being not believed by someone to have been bald.” What all this amounts to is a confusion between (i) rules of formation, and (ii) principles of individuation. They are not the same; and it is philosophically the same as to assume that an object may satisfy a predicate without its nature being modified. This cannot be the case with reference to properties; and this seems to be the natural way in which we understand ‘property’.

[30] As Wittgenstein would say, different variables; the apparentness of variables as they figure in quantification theory preoccupied him for far too long, and it figures too frequently in his correspondence with Russell.

[31] Hintikka (1).

[32] If quantification theory “had first been developed in connection with Geometry,” writes Hintikka, loc. cit., “it is not impossible that an exclusive interpretation of variables should have been adopted.” It is well known that Frege developed it in connection with arithmetic and set theory.

[33] Emphasis added.

[34] (2), Sec.67.

[35] (5), p.8, emphasis added. Does it emerge whether, for Frege, it is an explanation, or a definition, or an axiom, or something else!

[36] Identity provides conditions for substitutivity, and not the other way round. Let us call this Geach-criterion for it follows from the principle to be called Geach-principle, which he formulates in his Reference and Generality. According to it, the reference of an expression must be specific in some way that does not involve first determining whether the sentence in which it occurs is true. Geach holds - justifiably, I think - that all theories of reference which violate this principle are false theories. Incidentally, Wittgenstein passes the Geach-test, though it is likely that he, being a student of Wittgenstein, was informed by him.

[37] Emphasis added. I wonder why Wittgenstein gave his argument in epistemic terms creating the impression that the impossibility is epistemic. On the other hand what he is really up to is to show that it is logically impossible. Anyhow, this is not of much importance here.

[38] Emphasis added.

[39] I shall not push this point further because that would take me in a different direction and would divert me from the intended goal of this enquiry.

[40] The Hintikkas think that the logical form of an object is what can be known about it a priori. Though interesting in itself, this may not be of any use because all that can be known of an object a priori is that it is an object - its unity, that is. And this needs to be harmonized with Wittgenstein's thesis that there is no a priori truth, for it might be argued that nothing that is true of an object is known a priori.

[41] See my (8), where I argued that for Wittgenstein the reference of the semantic parts of a sentence determine its sense, this is to say that sense is a function of reference.

[42] Interpolation ours; see my (8).

[43] From the mere occurrence of the term 'essence' one should not smell neo-Thomism in Wittgenstein. In Tractatus most of the technical terms are duplicated; essence, nature, form, ... all mean the same thing.

[44] See my (4).

[45] Wittgenstein too uses the notion of "vorstellungen" with its psychological affiliations, but in a context which is befitting; see for instance his (3), 1.300 where he is trying to show how we can form images of sensations.

[46] The entry for 5.9.1914 in (5).

[47] The Hintikkas think that this has an additional advantage in the sense that it can be applied to sentences of the form "description-relation-name" as well as to sentences of the form "description-relation-description", or even "name-relation-name", and thereby enable us to overcome the limitations of Frege's system; see their Investigating Wittgenstein.

[48] In the sense that he sent to Russell a copy of the typescript which is the same as the present Tractatus; it even bore the present title, it is reported (though I cannot believe it).

[49] In March 1919 he wrote to Russell: "I finished the book in Aug 1918."

[50] This was questioned in 1961; see my (7).

[51] Specially (10).

[52] Nor did he stop thinking on the issues he earlier thought about; his apparent hibernation proved to be covert preparation for the future overt action.

[53] We can say that Ramsey's work on foundations is the result of his acceptance of what Russell aimed at in the Principia and Wittgenstein's criticism of the way in which Russell executed it.

[54] Mind, 1923.

[55] Foundations of Mathematics, 1925.

[56] ibid., p.179

[57] ibid., p.181

[58] ibid., <u>loc. cit.</u>

[59] ibid., 182

[60] ibid., p.179; perhaps Ramsey is referring to Wittgenstein's so-called "newthink".

[61] ibid., p.203; emphasis added. Note how Leibnitz's principle as I interpreted it in the Coda figures here. This is not to reject the principle but to see one of the consequences of the principle as Leibnitz seems to have taken it. Ramsey was attempting at it perhaps without being aware of what he was doing. It could even be that he was informed by Aristotle's notion of qualities, perhaps indirectly; it was the time when Stout was articulating his notion of particulars with which Cambridge philosophers were familiar. All this amounts to saying: this is this red, and that is that red. 'This red' and 'That red' - red (1) and red (2) - are, for Wittgenstein, not further analyzable at all. The reason he gives for this is based on something similar to his rejection of the view that a universal sentence is a product of singular sentences. As he remarked in (2): "One might think - and I thought not long ago - that a statement expressing the degree of a quality could be analyzed into a product of single statements of quality and a complementing supplementary statement I maintain that the statement which attributes a degree to a quality cannot be further analyzed."

[62] <u>Foundations of Mathematics</u>, 1925, p.201.

[63] Ramsey formulates it in terms of <u>equality</u>; this further complicates the matter. Given the extensionalist thesis that two sentences are equivalent when they have the same truth-value, another needless problem comes into being in situations where x is correlated with the sentence S, and y not with S but with some truth-functional equivalent of S. The way I formulated it circumvents that problem while retaining the essential content of Ramsey's message.

[64] Emphasis added; juxtapose this with Russell's view that a sentence where = is flanked by genuine proper names is either a tautology or a contradiction depending upon whether it is flanked by the same or different names. It amounts to saying that if $x = y$, then $L(x = y)$, and if $x \neq y$, then $L(x \neq y)$ or is it $\sim L(x = y)$! I shall soon consider Wittgenstein's objections to this. All this may throw some light on the Barcon formula, and Kripke-necessity; they seem to go deeper than it is usually thought, and seem to spread wider than it is usually assumed.

[65] Of course, Ramsey noted the difference between the two. $(\phi)(\phi x \rightarrow \phi y)$ is a tautology when ϕ is a predicative expression <u>and</u> $x = y$, but it is not a contradiction when $x \neq y$.

[66] But in (15) Wittgenstein remarked that Ramsey was a bourgeois in the sense that he cannot think of alternatives to the received; and for Wittgenstein this inability is equivalent to spiritual poverty. See his (14).

[67] Part II, Sec.16, p.375.

[68] In Foundations of Mathematics Ramsey uses 'definition' and 'explanation' synonymously; see for instance p.203 where he talks of defining or explaining propositional functions in extension.

[69] Waismann (1), p.189.

[70] pp.190-191.

[71] Foegelin (1).

[72] This is a short for $(\phi_e)(\phi x \equiv \phi y)$.

[73] Wittgenstein's attitude towards numbers is almost Kroneckerian, perhaps even Promethean.

[74] (5), entry for 22.5.1915; emphasis added except in the first place.

[75] (5), entry for 4.5.1915.

[76] Compare it with Brouwer's understanding of "spread" as a pair of laws governing the generation process under which its elements grow, in order to note how Wittgenstein was thinking "intuitionistically" though his acquaintance with Brouwer's thoughts was still in future. The results of the application of a rule cannot be determined prior to the application itself. Wittgenstein's objection to Russell's theory of types is based on this very idea. As he remarks in 3.331 of the Tractatus:

> It can be seen that Russell must be wrong, because he has to mention the referents of signs when establishing rules for them.

[77] Included in (10), see Part II, Sec.16.

[78] I wonder whether his not paying any attention to Moore's Principia Ethica is due to its having as its motto Butler's profundity! It is universally admitted that he is a man of strong likes and dislikes.

[79] Foegelin gives in (1) an interesting explication of the passage below from Wittgenstein's (3):

> 'A thing is identical to itself' - There is no finer example of a useless proposition.

As it is not strictly relevant to the point I want to make here, I shall not go into details either about that passage or about Foegelin's explication. I am restricting my attention to problems that arise in relation to representation; and Wittgenstein made that remark in connection with the notion of following a rule, which, though important, falls outside the scope of my concern here.

[80] See my (6).

[81] Respectively p.68 and p.317.

[82] This explanation might sound metaphysical. But Wittgenstein is not averse to metaphysics; of course, he too, like all of us, was contemptuous of bad metaphysics. If in 1913 he recorded in (5) - see Appendix-I:

"philosophy consists of logic and metaphysics; the former is its basis,"

during the late forties he is reported to have told Drury, as he recorded in his The Danger of Words:

> "Don't think that I despise metaphysics or ridicule it. On the contrary, I regard the great metaphysical writings of the past as among the noblest productions of the human kind."

[83] (15), p.61e; this passage is of 1947 vintage.

[84] Emphasis added; Mcguinness (see Copi and Bird (eds), Essays on Wittgenstein's Tractatus) offers an alternative and interesting reading of this passage. Normally it is understood that Wittgenstein is making use of the criterion he gave in 4.461 - 4.4611 of the Tractatus to distinguish the class of sentences without sense (sinnlos). Mcguinness points out how in 5.5303 itself the criterion is alternatively stated and how sentences without sense (sinnlos) are contrasted with sentences which are meaningless (unsinning). But, as he sees it, this leads to a difficulty, because "it implies that, where as a and b are two things. 'a = b' is not a contradiction (otherwise by 4.4611 it would not be an "unsinn"); but if so, it would seem that 'a = a' is not a tautology ... if, however, 'a = a' is not a tautology, why should it be an unsinn?" I trust that my clarification of the parenthesis in 3.203 goes at least half the way to set things aright. I agree with Mcguinness that here there is an ambiguity, but in the light of my clarification of Wittgenstein's stand on definite descriptions, I think that it is unwarranted to come to the conclusion, as Mcguinness does, that "Wittgenstein ought to have said that ... (=) ... did lead to sense when put between expressions one of which was a description, but would lead to unsinn when placed between names.:

[85] I am not suggesting that he assiduously studied Leibnitz's works; Leibnitz's name does not figure in the list of the persons who influenced him which he gave in (15). But he went through Russell's work on Leibnitz, and it is likely that he saw Leibnitz-Clarke Correspondence published from Manchester. In any case, coming as he does from a haute bourgeoisie family of Austria it is unlikely that he was ignorant of the ideas of Leibnitz who was without doubt one of the greatest minds that Europe produced and is likely to produce.

[86] Russell's (1), p.54.

[87] Leibnitz's (1), Vol.2, p.1147.

[88] Here we already have intimations of the axiom of reducibility. There are sound reasons for entertaining that Wittgenstein took Leibnitz's notion of properties seriously and incorporated it in the Tractatus. That alone permits him to turn properties into individuals and then fit them into the frame according to which facts are concatenations of impenetrable objects. Individuals and individual properties both are objects and figure in the logical discernibilia of the world. It is well-known that according to the creation myth with which the Tractatus opens, these discernibilia are facts; see my (1), Ch.1. In his (3), Part I, p.46 Wittgenstein cites Plato's Theatetus approvingly to say that

> "it is impossible to give an account of any primitive element ... (for) ... its name is all that it has."

He admits that Russell's individuals and his objects "are such primary objects."

Bibliography

Black, M.

1. *A Companion to Wittgenstein's "Tractatus"*, Routledge and Kegan Paul, 1964.

Diamond, C., (ed)

1. *Wittgenstein's Lectures on Foundations of Mathematics* - Cambridge 1939, Cornell University Press, 1976.

Foegelin, R., J.,

1. *Wittgenstein*, Routledge and Kegan Paul, 1976.
2. Wittgenstein on Identity, *Synthese*, 1983.

Frege, G.,

1. *Begriffsschrift und andere Aufsatze* (ed) Angellelli, I., Olms, 1964.
2. *Die Grundlagen der Arithmetic*, 1884; English translation by Austin, J.L., *The Foundations of Arithmetic: A Logico-Mathematical Enquiry into the Concept of Number*. Blackwell, 1960.
3. *Grundgesetze der Arithmetic*, Vol. I, 1983; English translation by Furth, M., *The Basic Laws of Arithmetic - Exposition of the System*, University of California Press, 1964.
4. *Uber Sinn und Bedeutung*, 1892; English translation by Black, M., and Geach, P.T., (eds) *Translation from the Philosophical Writings of Gottleib Frege*, Blackwell, 1952.
5. *Review of Hussrel's Logische Untersuchungen*, included in Black and Geach (eds) cited above.
6. *Philosophical and Mathematical Correspondence*, English translation by Kaal, H., Chicago University Press, 1980.

Hintikka, J.,

1. Identity, Variables, and Impredicative Definitions, *The Journal of Symbolic Logic*, 1956.

Leibnitz, G.F.,

1. *Philosophical Papers and Letters*, (ed) Leomker, L.E., Chicago University Press, 1956.

Muehlmann, R.,

1. Russell and Wittgenstein on Identity, *The Philosophical Quarterly*, 1969.

Ramsey, F.,

1. Foundations of Mathematics, 1925, included in Braithwaite, R.J., (ed) *Foundations of Mathematics*, Routledge and Kegan Paul, 1931..

Rao, A.P.,

1. *A Survey of Wittgenstein's Theory of Meaning*, Indian University Press, Calcutta, 1965.
2. *Aristotle's First Philosophy in Proper Perspective*, Center of Advanced Study in Philosophy, Santiniketan, 1968.
3. The Concept of Logic, *Notre Dame Journal of Formal Logic*, 1973.
4. Towards a Free Description Theory, *Notre Dame Journal of Formal Logic*, 1975.
5. On the Indiscernibility of Facts and Propositions, *Indian Philosophical Quarterly*, 1976.
6. On Translating 3.203 of *Tractatus*, *Jadvapur Studies in Philosophy*, Vol. 5, Macmillian, 1984.
7. A Survey of Free Logic, *Modern Logic*, April 1996.
8. Wittgenstein - A Second Look I, II, and III, *Journal of the Indian Council of Philosophical Research*, 1985 & 1986.
9. A Neglected Chapter in the History of Logic, (forthcoming).

Russell, B.,

1. *A Critical Exposition of the Philosophy of Leibnitz*, Allen and Unwin, 1900.
2. *Principle of Mathematics*, Allen and Unwin, 1901.
3. On Denoting, *Mind*, 1905.
4. *Principia Mathematica*, Vol. I, Cambridge University Press, 1910.
5. *An Introduction to Mathematical Philosophy*, Allen and Unwin, 1919.
6. Introduction to Wittgenstein (1), 1922.
7. *My Philosophical Development*, Allen and Unwin, 1957.

Strawson, P.F.,

1. *Philosophical Logic*, Oxford University Press, 1967.

Waismann, F.,
1. *Wittgenstein and the Vienna Circle*, Blackwell, 1967.

White, R.,
1. Wittgenstein on Identity, *Aristotelian Society Proceedings*, 1978-79.

Williams, C.F.,
1. Is Identity a Relation? *Aristotelian Society Proceedings*, 1981.

Wittgenstein, L.,
1. *Tractatus Logico-Philosophicus*, (1922), with an introduction by Russell, B., 1922; second translation by Pears, D., and Mcguinness, B., Routledge and Kegan Paul, London, 1961.
2. Some Remarks on Logical Form, *Aristotelian Society Proceedings*, 1929.
3. *Philosophical Investigations*, translated by Anscombe, G.E.M., Blackwell, Oxford, 1958.
4. *Blue and Brown Books*, edited with a preface by Rhees, R., Blackwell, Oxford, 1969.
5. *Note Books 1914-1916*, translated by Anscombe, G.E.M., Blackwell, 1961.
6. *Letters From Ludwig Wittgenstein with a Memoir*, translated by Furtmuller, Blackwell, 1967.
7. *On Certainty*, translated by Paul, D., and Anscombe, G.E.M., Blackwell, 1969.
8. *Ludwig Wittgenstein - Letters to Russell, Keynes, and Moore*, edited with an introduction by von Wright, G.H., Blackwell, 1974.
9. *Remarks on Colour*, translated by Noalister, L.L., and Shattle, M., Blackwell, 1977.
10. *Philosophical Grammar*, translated by Kenny, A., Blackwell, 1978.
11. *Proto-Tractatus*, edited by von Wright, G.H., Mcguinness, B.F., and Nyberg, T., Routledge and Kegan Paul, 1971.
12. *Remarks on the Foundations of Mathematics*, translated by Anscombe, G.E.M., Blackwell, 1978.
13. *Zettle*, translated by Anscombe, G.E.M., Blackwell, 1979.
14. Remarks on Frazer's Golden Bough, translated by Bersluis, J., (in) *Wittgenstein - Sources and Perspectives*, edited by Luckhardt, C.G., Harvester Press, 1979, pp.61-81.
15. *Culture and Value*, translated by Winch, P., Blackwell, 1980.

Index

Notes and references to Russell in Part I and to Wittgenstein in Part II are not indexed.